WEST SIDE RISING

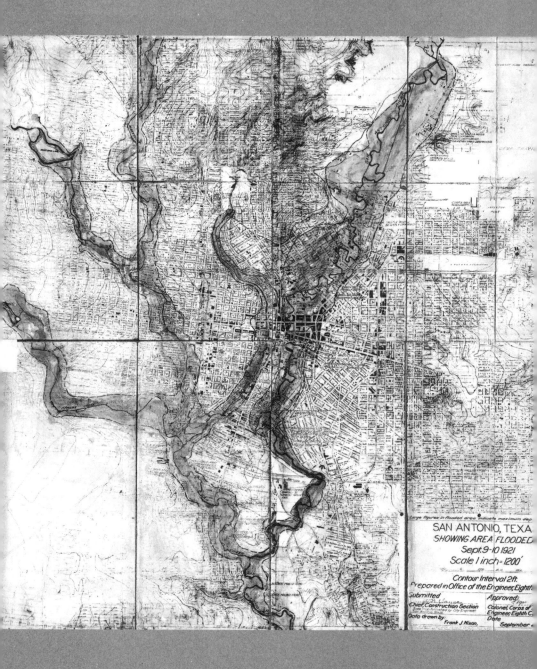

SAN ANTONIO, TEXAS
SHOWING AREA FLOODED
Sept. 9-10 1921
Scale 1 inch = 1200'

Contour Interval 2 ft.
Prepared in Office of the Engineer, Eighth

Submitted   Approved;
Chief, Construction Section   Colonel, Corps of
                              Engineer, Eighth C.
Data drawn by   Date
            Frank J. Nixon,   September

WEST SIDE RISING

*How San Antonio's 1921 Flood Devastated a City
and Sparked a Latino Environmental Justice Movement*

CHAR MILLER

Foreword by
JULIÁN CASTRO

MAVERICK BOOKS / TRINITY UNIVERSITY PRESS
SAN ANTONIO, TEXAS

Maverick Books, an imprint of
Trinity University Press
San Antonio, Texas 78212

Book design by BookMatters
Cover design by Rebecca Lown

Cover: Children gathered at a US Army food wagon in the aftermath
of the 1921 flood (detail). Reproduced by permission from the Library
of Congress, Prints and Photographs Division, American National Red
Cross Collection.

ISBN 978-1-59534-973-6 paperback
ISBN 978-1-59534-939-2 ebook

Trinity University Press strives to produce its books using methods and
materials in an environmentally sensitive manner. We favor working
with manufacturers that practice sustainable management of all natural
resources, produce paper using recycled stock, and manage forests with
the best possible practices for people, biodiversity, and sustainability.
The press is a member of the Green Press Initiative, a nonprofit program
dedicated to supporting publishers in their efforts to reduce their
impacts on endangered forests, climate change, and forest-dependent
communities.

The paper used in this publication meets the minimum requirements
of the American National Standard for Information Sciences—
Permanence of Paper for Printed Library Materials, ANSI 39.48-1992.

CIP data on file at the Library of Congress

25 24 23 22 | 5 4 3 2 1

⊣ CONTENTS ⊢

Julián Castro

In the early 1920s, when my grandmother immigrated to the West Side of San Antonio from northern Mexico as a seven-year-old orphan, signs reading "No Dogs, Negroes, or Mexicans Allowed" greeted her on store-front windows. Racism against the Latino community was conspicuous, but it was also systemic, baked into the decision-making of every powerful local institution. Indeed, a century before the term "systemic racism" leapt onto the placards of protestors demonstrating against racial injustice in America, San Antonio civic leaders were well practiced at it, consigning the mostly Mexican and Mexican American residents of the city's West Side to a second-class existence. In *West Side Rising*, Char Miller chronicles the deadly consequences that ensued when that bigotry made West Side residents particularly vulnerable to nature's fury during the Great Flood of 1921, which resulted in dozens of deaths and the widespread destruction of property in the barrio, and in the years that followed as subsequent flooding claimed more lives and property.

As a child of the West Side in the 1970s and 1980s, I became familiar with the marginalization Miller describes—unpaved roads, shoddy drainage, chronically underfunded schools, meager opportunity. But I also saw another side of the story, a more hopeful side: West Siders had been disempowered, but they were not helpless. Time and again, they empowered themselves through community organizing and steadily made inroads.

To his credit, Miller wisely avoids the temptation to paint the marginalized as simply victims and instead details how they fought back. In the aftermath of the 1921 flood, they helped each other through mutual aid societies like Cruz Azul and the League of United Latin American Citizens. They propelled trailblazing public servants like congressman Henry B. González into office to fight successfully for more resources from Washington, DC. They showed up at the ballot box to make important structural changes to local government, including the creation of single-member city council districts, that massively boosted the resources allocated to the West Side. And they organized barrio residents through Communities Organized for Public Service and other community-based efforts to demand infrastructure improvements and pursue environmental justice. Through Miller's exploration of these efforts, we learn once more the power of everyday people to affect change.

Miller is a meticulous researcher and lively storyteller whose work is essential reading because its lessons are relevant far beyond the experience of San Antonio's Latino community, and because those lessons have never been more timely. We are reminded that policy-making is about choices, so we should choose wisely. In the years after the 1921 flood, San Antonio city leaders invested millions of dollars to protect the lives and property of the mostly white downtown business owners. Building the Olmos Dam in Alamo Heights in 1926 was their answer. They neglected the infrastructure needs of the Latino neighborhoods, even though these had been hit hardest by the storm.

I represented several West Side neighborhoods as a young councilman in the early 2000s. I can remember many visits with homeowners after a heavy rain. Most often they were elderly and had lived in their modest bungalows for decades. They took me inside with a mix of anger, sadness, and incredulity. "¡Mira!" They would point out stormwater pooling in a corner of their kitchen or garage. They had been waiting decades for the city government to fix the drainage problem on their streets, they told me. I soon realized that, even if I dedicated every single penny of my district's infrastructure budget solely to address their needs, it wouldn't be enough to catch up after decades of underinvestment in the West Side. Twenty years later, in 2021, I have no doubt many are waiting still.

Just about every American city has a similar tale to tell. From "Cancer Alley," the eighty-five-mile stretch of the Mississippi River in Louisiana jammed with oil refineries and petrochemical plants, to the recent management failure of the water system in Flint, Michigan, environmental racism is all too common. People of color have been marginalized by government and industry in communities across our country, and they have pushed back, in strikingly familiar ways, to improve their quality of life. The fight for environmental justice continues.

Recently President Joe Biden proposed a historic $2 trillion infrastructure bill, the largest single investment in infrastructure in our nation's history. What a perfect time to learn the lessons *West Side Rising* can teach us.

"CULEBRA DE AGUA"

The keys were a clue. Something was amiss.

Still inserted in the front door of Ben Corbo's popular fruit market at 422 Saint Mary's Street, they attracted the attention of employees of the Hughes Auto Livery nearby. The staff had been in the process of clearing out their shop in the aftermath of a horrific flood that had torn through San Antonio, Texas, two days earlier.

It was 1921. The flood had begun on the evening of Friday, September 9, with heavy rain lashing the San Antonio River watershed; by 10 p.m. Saint Mary's Street had become a river. Within two hours raging waters were crashing through downtown and devastating portions of the West Side neighborhood where the Corbo family lived. Some of the auto livery's personnel, as they had headed home through the whipping winds that evening, stopped by the fruit market to check on Corbo. He asked the men to escort his son to the family's home on Monterey Street, one and a half miles west, while he remained "in the store a few minutes longer in an attempt to save some of the property."[1]

That was the last anyone had heard from him. The flood took out power and telecommunications for two days, and the city imposed a strict curfew that blocked access to the badly damaged downtown. It was not until Sunday morning, September 11, that some of the livery staff workers returned to inspect their shop. As they swung past Corbo's market, they spied his keys.

Fearing the worst, the men sprang into action: "Hammering their way through the front of the building, rescuers attacked the wreckage…and after a search that continued into the middle of the afternoon the body was located beneath debris that filled the rear of the building."[2]

{ 2 } Those minutes trying to save some of his property cost Ben Corbo his life. In one sense, his death was preventable. Had he left the shop with his son, he might have lived through that terrifying night; his family managed to survive. But then history is replete with what-ifs, questions that have no answers yet paradoxically point the way to some explanatory patterns. The fruit vendor, after all, was like many others who died in the most fatal flood in San Antonio's history. Swirling with street pavers, automobiles, furniture, and branches, turbulent stormwaters undercut houses, commercial buildings, and bridges, and killed many who lay asleep in their beds. Other residents, who had managed to escape their battered dwellings, were sucked into the maelstrom. Corbo was not alone in being trapped by the relentless force of the waterborne debris.

Yet his demise—he was one of an estimated eighty who succumbed that night—was unusual in this respect: only four people perished as a result of the San Antonio River's floodwaters, and Corbo was the only adult. The other three were young children the river pulled from their parents' arms. The vast majority of those who perished were on the city's densely populated West Side, in an area known locally as the corral or jacal district (so named for the huts and shacks many of its residents occupied). These rough shelters were no match for the powerful floodwaters that raced down the West Side's interlacing of creeks—the Alazán, Martínez, Zarzamora, Apache, and San Pedro. That evening, the Alazán, which curved one block west of Corbo's home, proved the deadliest.

This disparity in the demographics and distribution of death in San Antonio dovetails with a statewide pattern with regard to the disaster: spatial inequities, ethnic discrimination, and environmental injustices determined who survived and who died in this massive flood. "The total number of lives lost will never be known," wrote US Geological Survey water engineer C. E. Ellsworth in an extensive analysis of impact of the 1921 flood across central Texas, "but the best estimates available indicate that at least 224 people were drowned, most of whom were Mexicans

Some of the dozens of homes along Alazán Creek that were washed away.

who lived in poorly constructed houses, built along the low banks of the streams." His careful qualification of the number of fatalities across the Lone Star State held true as well for San Antonio. "Undoubtedly many others were drowned and never reported missing. Many bodies were carried miles and buried in sand, mud, and debris along the river bottoms."[3]

A particularly mournful example came to light one month after the flood. Several days before the storm, Mariano Escobedo, who lived with his wife, Maria, and two children in a shack on the banks of the Alazán near El Paso Street, had left town to find work in West Texas. It was some time before he heard about the flood, and even longer before he was able to scrape together enough money to return to San Antonio. "Persons in whom he applied for aid doubted his story and refused to help him," so Escobedo saved "every penny he could" and gradually worked his way back to town. His small abode, which was located "in the path of the torrent that swept down Alazan Creek," had vanished. Friends and acquaintances had no news of his family, and his dogged search for clues close to home and expanding downstream came up empty. The Red Cross, which "assisted him in every way possible," had no record of Maria, Josephine M. Escobedo (age seven), or Jesús M. Escobedo (four months). The city police suspected that their bodies "may have been carried many miles away by the flood waters" and as a result probably would never be found. They also told the *San Antonio Express* that the disappearance of the three Escobedos was not unusual: "There were numbers of instances of persons in the Mexican district along the Alazan Creek washed away that were not brought to the attention of the Red Cross or other officials."[4]

Those who perished in the floods that year—many of Mexican heritage, poor men and women whose manual labor was seasonal and low-paying, and who therefore settled in the least expensive and most flood-prone terrain—were made all the more vulnerable because of the region's geography, topography, and climate. The most devastated communities all hugged the Balcones Escarpment, a geological fault zone that runs for roughly 450 miles. Like a lopsided smile, it curves east and north from Del Rio on the Rio Grande all the way to the Red River delineating the Texas-Oklahoma border; it forms hilly contours that define San Antonio, New Braunfels, San Marcos, and Austin, and from Austin north to Georgetown and beyond. A modern signifier is Interstate 35, which parallels the escarpment to its east from San Antonio north.

This rumpled and craggy landform is critical in several respects. It demarcates the southern terminus of the Great Plains, an elevated terrain that in Texas is known as the Edwards Plateau; some sections of the rough limestone and thin-soiled landscape are as high as two thousand feet or so. The land that slopes away from the fault line toward to the Gulf of Mexico is the southern coastal plain. The reciprocal relationship between these two masses and longstanding climatological patterns can produce wild swings in local weather. Because this region falls loosely within the transition zone between the humid eastern section of the United States and the arid West, the climate can toggle between deluge and drought, an oscillation fueled in part by whether an El Niño or La Niña system prevails. Another trigger mechanism involves the Gulf of Mexico. If its bathwater-warm, moisture-laden air pushes onshore, the flow will slowly lift with the rise in elevation. Once it reaches the escarpment the uplift is more abrupt, forcing the moist air to interact with colder temperatures above. The moisture condenses, the cooler air falls, then it is warmed and rises again, a cycle of convection that can result in major thunderstorms. "It has long been recognized," notes C. Terrell Bartlett in a contemporary

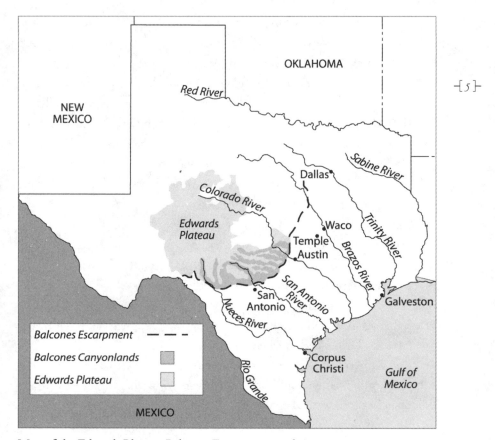

Map of the Edwards Plateau, Balcones Escarpment, and river systems.

assessment of the 1921 flood, "that in many cases the sudden rise along the Balcones Escarpment causes intense precipitation along and just above its margin." This intensity can generate blockbuster floods, a reality that has led the National Weather Service to dub San Antonio and the larger region along the escarpment "Flash-Flood Alley."[5]

Over the millennia, rains have carved a series of creeks, streams, and rivers into the limestone that widen as they reach the coastal plain and head to the Gulf, an erosive process that flooding could accelerate. In their more placid state, the San Antonio River, as well as the Guadalupe (New Braunfels), Comal (San Marcos), and Colorado (Austin) Rivers, have attracted and sustained generations of indigenous communities and in time Spanish, Mexican, and American settler-colonists. Each group took advantage of these systems' life-sustaining properties and the ecological abundance that came from living within the fertile intersection of different biozones, prairie and plateau, grassland, riverine, and woodland. But should a furious thunderstorm explode over the upper reaches of the local watersheds, the floodwaters that sluiced down innumerable gullies and ravines and surged into streams could have devastating consequences. Within moments, the almost dry San Antonio River and its arroyo-like tributaries could become churning torrents.

That is what happened on September 9 and 10, 1921: San Antonio went underwater. So did New Braunfels, San Marcos, and Austin, along with the smaller communities of Taylor and Thrall. Each scouring event profoundly altered the communities. In this instance, the staggering volume of water that came crashing down when the storm broke and the skies opened up was the result of a slow-moving tropical depression that several days earlier had crashed ashore in northern Mexico. As it spun over the Rio Grande, the system dumped upward of 6 inches on Laredo, submerging low-lying neighborhoods. Pressing north, it cycled along the Balcones Escarpment, where storm cells unleashed their full fury. Thrall recorded an eye-popping 38.21 inches in a twenty-four-hour period, in what was believed to be one of the largest single-day rainfall events in the continental United States. (For the record, Alvin, Texas, received a swamping 43 inches in 1979, courtesy of Hurricane Claudette.) More than 23 inches fell on Taylor, and Austin got 18.23.[6]

These fixed-site numbers do not tell the whole story, of course, because upstream from each of these communities, heavy rains were coming down over the folds of the Edwards Plateau and surging into the region's major and minor river systems. These roiling waters swept over farms, ranches, and feedlots in short order; houses and barns were pushed off their foundations and careened downstream; uprooted trees became battering rams, slamming into buildings, tearing up transportation infrastructure, and collapsing bridges. Property damage was severe, but the number of fatalities was even more so: in Taylor, eighty-seven people died, and another six perished in its surrounding county of Williamson. Six people drowned in Travis County, where Austin is located. By all measures, the 1921 flood was the most devastating in the history of the Lone Star State.[7]

Nowhere was this truer than in San Antonio, a city whose eighteenth-century Spanish planners had platted in a floodplain so that its streets, plazas, and residential areas lay within the embrace of two waterways, the San Antonio River on the east and San Pedro Creek to the west. As a result, the community periodically had foundered: until its roads were hardened in the late nineteenth century, even light rains had turned them into a muddy morass. The foundations of new upper- and middle-class housing built after the mid-nineteenth century often were lifted above street level in hopes of keeping their occupants clear of moderate flooding (a strategy that had some success). But nothing protected the central core during the floods of 1819, 1845, and 1865, when so much water poured down river and creek that their combined overflow merged, trapping residents and devastating homes, churches, and commercial properties. Two punishing floods in 1913, and another one in 1914, destroyed parts of the city, especially the West Side. Like the most damaging of all, that of September 1921, these floods added to the pattern of inundation that was built into San Antonio's environmental history.

Floods made (and unmade) the San Antonio region over time. Their impact can be read in the contours of the shape-shifting structure of the river and valley in which the city is located, the composition of its soils, the health of its riparian habitats, and the resilience of the countless species that over thousands of years have been drawn to the place. It has been a rich land in part because of the floods that carried silt and other material down to the valley that added to its biological health. The indigenous peoples of the region had managed and utilized this biodiversity in the vicinity of their semipermanent settlements close to local springs and streams. The Spanish entradas took careful note of these same environmental riches and the human presence in the place that they would name after Saint Anthony of Padua.[8]

What the Payaya and other tribal groups knew, and the Spanish quickly would discover, was that the San Antonio valley lies within a web

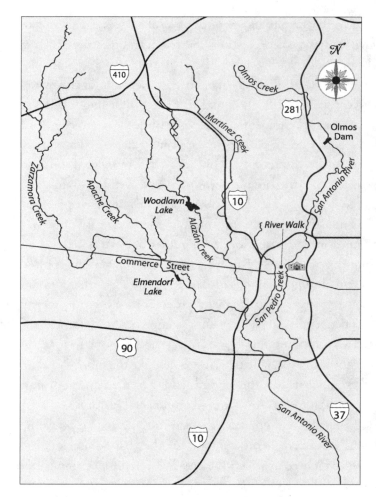

Map of the San Antonio River and the West Side creeks.

of rivers, streams, and creeks. Imagine a handprint, with San Antonio as the palm. Like fingers, five major waterways define the riparian relationship between the low hills to the north and the flatlands below. From east to west, these systems include Cibolo and Salado Creeks, Olmos Creek and the San Antonio River (now described as the upper and lower watersheds of the San Antonio River), Leon Creek, and the Medina River. Each is fed by the crystal-clear waters of the Edwards Aquifer that bubble up from its springs, bogs, and wallows; each ultimately merges with the San Antonio River. Inlaid within these systems, particularly on the community's west side, is a set of small streams. The shortest of these, San Pedro Creek, is spring fed; in dry weather, Alazán, Martínez, Apache, and Zarzamora Creeks carry minimal flow, functioning as dry ditches. When rain falls, they can rise quickly, powerfully, and fatally.

These floods have also left their (water) mark in the historical record. One melancholic ledger can be seen in local cemeteries, especially the historic city cemeteries on the East Side and San Fernando Cemetery on the near West Side—the latter of which is where most of those who died in the 1921 flood were interred. Gravestones bearing the dates of death offer silent reminders of the sudden demise of those who fell victim to one of the many inundations that have rampaged through the city since the eighteenth century.

Oral histories, by contrast, often speak of and for those who survived, and the fact that they are recalled, written down, or preserved in archival form may indicate the class and experience of their narrators. At least in retrospect, for example, Fred Groos's recollection of the 1921 flood was that it seemed something of a lark. That perception may have been aided by his age—he was ten when the flood struck. It helped, too, that he lived in a large, well-anchored and framed house in the King William district, a reflection of his family's elevated status and its ability to afford a home that could withstand the floodwaters that routinely coursed through the river-abutting neighborhood south of downtown. Awakened by his mother, he watched from the second-story, wraparound porch as "the water came up, now from all sides." His grandfather, suspecting that the family needed to get higher up, propped a stepladder on the porch to a nearby elm tree overhanging the house. "By now the water was still rising so all but grandfather, dogs included, climbed the ladder and went to the roof." Groos's memory was that the water rose to within four inches of the second story, but that it had "pretty well passed on by early morning and only the mess, in our case, the utter devastation in most cases, was left." His family's good fortune, he knew, was relative.[9]

His insight about the disparity of loss was revealed in an interview a

San Antonio Express reporter conducted with an unnamed pair of elderly West Siders who lived just a few miles west of the Groos household. While surveying the intersection of Lakeview Avenue and Alazán Creek, the journalist spotted a "small Mexican shack [that] stood out prominently near the bed of the creek, while other homes were swept away by the water. The shack was occupied by an aged Mexican couple who while unable to speak good English explained that their entire household belongings consisting of a large dry goods box, in which they slept, a table, an oil stove, and dozens of tin cans, which were used as cooking utensils, were carried off by the water."[10]

Though not as devastating as that couple's night, Dulce Watson's was bad enough. She and her husband, Robert, lived in a brand-new bungalow on the corner of Natalen Avenue and Margaret Street, one block east of River Avenue and Brackenridge Park. Out of town on a business trip, her husband would spend the night in his car on high ground trapped between two roaring creeks. His wife's situation was far more dangerous, a threat she sensed in the middle of the night when, nearly eight months pregnant with her third child, she got out of bed to go to the bathroom—and stepped into ankle-deep water. "The sound of the rain was a noisy, incessant battering, and garbage cans and other debris rushed along in an angry current, hitting the telephone pole in the street or bumping against the house." Dulce quickly grabbed her two young sons and climbed onto a four-foot-high pony wall separating the kitchen and breakfast nook. With her eldest sitting beside her and her nineteen-month-old nestled "in what remained of her lap," she watched the water rise until it overtopped the breakfast table. As the floodwaters slowly receded in the early morning hours, and with her children hungry, she scrounged for shoes to protect her feet. Finding none wearable, she walked barefoot in the "slimy ooze." She slipped and dropped a glass baby bottle she was carrying; it shattered and she fell on its shards, suffering deep cuts to her hand and wrist that just missed the radial artery. The scars were "visible as long as she lived." Rescued by her father, but in shock, and deeply distressed that her husband might have been killed, Dulce was sedated. Over the years, her daughter recounted, the story was frequently told within the family

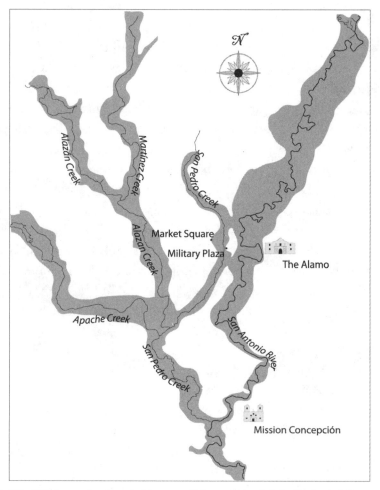

The US Army developed this map of the 1921 floodplains and rainfall amounts for its post-flood report.

circle, for "no other event in her life matched the horrendous experience of that night."[11]

Luck, no less than racial and ethnic identity and social standing, mattered in these and other experiences. A ladder was close at hand in one. The roiling water pulled in one direction but not another. In a third, a well-placed telephone pole intercepted and shattered into "smithereens" a garage racing down Margaret Street that would have crashed into Dulce Watson's cottage with unknown consequences. The materiality of these objects, like Ben Corbo's keys, adds weight to the narratives.

By the same token, the absence of objects may limit what can be recovered and thus what and whose story can be told. Consider the reams of legal, ecclesiastical, and governmental records that floodwaters have turned into a sodden mass or sucked out of offices, churches, and courthouses and sent downriver like so much detritus. Still, some paper trails remain: copious amounts of ink have been spilled to describe what floods have wrought. The damage has inspired journalists, editors, engineers, politicians, and residents to put words on a page to describe and debate how best to control what had seemed so uncontrollable. Even this book, which draws on these written accounts, testifies to, and surely is not the last accounting of, the central role floods have played in defining how a community has spoken of its travails.

The built landscape is not without a voice, either. Tracing the homes that have endured or been torn down, the commercial blocks and business edifices that have withstood inundation, been pumped out and repaired, or newly constructed in a flood's wake, is as telling as searching for the mills and diversion dams that are no more. Another way to trace the fate of infrastructure such as roads, bridges, and trestles that have weathered an overflow (or not) is to read city council minutes, though those records can both clarify and obscure what happened. Add to these documents the post-flood recycling of debris—lumber, steel, and brick—that has shored up storm-damaged structures. The past can be animated by the inanimate.

West Side Rising's overarching goal is to bring the history of the 1921 flood to life. It makes extensive use of a constellation of primary and secondary sources, including correspondence, government documents and photographs, oral histories, and newspaper accounts, as well as scholarly analyses of this and other floods, to establish a wider context to understand what happened in San Antonio. Many of these documents have been overlooked and are, as a result, all the more riveting. What they and other sources make clear is that this disaster, like so many others, was

not "natural," a calamity beyond human imagining and control. Quite the opposite. The death and damage of September 1921 were a direct consequence of a series of deliberate and very human decisions to build this city inside a floodplain and to rebuild there every time it foundered. San Antonio's actions mirrored those of the marshland cities to its east, Houston and New Orleans, and those more arid communities to its west, El Paso, Tucson, Phoenix, and Los Angeles. Frequently pummeled by floods, these cities' residents rebuilt where they had lived and even encroached deeper into local bayous and floodplains, adding to the misery of the next inundation. This nationwide set of responses reinforces historian Ted Steinberg's argument that disasters in the United States are produced "through a chain of human choices and natural occurrences."[12]

These choices, when read through the lens of race, class, and environmental justice, also reveal why particular people inhabited particular sections of the city. Like its southwestern peer cities, San Antonio was deeply divided and segregated, with a physical separation that found verbal expression. Its West Side residents were uniformly called "Mexican"—by whites, at least. "In Texas, the word Mexican is used to indicate the race, not a citizen or subject of the country," James L. Slayden, the city's former congressional representative, explained approvingly in January 1921, a rhetorical device that reinforced the subordination of those he labeled an invading "alien race."[13] His racist trope was an explanation for and linguistic evidence of the spatial segregation of Mexicans and Mexican Americans in El Paso, Phoenix, and Los Angeles, as well as San Antonio's West Side. Since the 1850s, this quadrant had housed the city's poorest residents who lived in substandard housing and experienced frequent outbreaks of tuberculosis, yellow fever, and other diseases. It was not an accident, then, that the September 1921 flood was especially lethal and catastrophic for those who were geographically marginalized.

There was nothing accidental about the reality that this impoverished community of color, like so many others elsewhere, routinely and disproportionately found itself in harm's way. It was the consequence of the persistence of conditions that environmental justice scholar Rob Nixon aptly describes as "slow violence," the systemic result of racial capitalism,

{ 13 }

spatial inequality, and political disenfranchisement. It was not accidental that San Antonio's power elite moved swiftly post-flood to distribute
charity to the beleaguered and then agreed to tax the community writ

large to fund the construction of a flood retention dam that secured only
the fortunes of the downtown commercial district. Using their manifold
resources, the city commissioners and the flood prevention committee,
the latter of which consisted of major downtown property owners and
leading engineers, developed a political consensus around the pressing
need for a dam and other related flood control infrastructure to protect
the business district's property values and boost the urban economy. The
strategy paid off, at least for them, but once again the West Side was
largely excluded from its benefits, a national pattern that dovetails with
Steinberg's generalized assessment of official responses to disaster. These
have, he writes, "contributed to a continuing cycle of death and destruction and also normalized the injustices of class and race."[14]

West Side Rising begins to examine these and other contexts by pairing
the prologue with chapter 1. The former offers a distressing account of
the massive 1819 inundation, sparked by a *culebra de agua*, a violent thunder-
storm-cum-flood that shattered the New Spain village and left it reeling.
Within a couple of decades, the catastrophe had largely been forgotten.
It seems to have completely disappeared from the community's collective
memory when a little more than a century later the 1921 flood ravaged
the city; that devastation is the subject of chapter 1, which recounts that
night of terror and provides a close analysis of who died, where, and why.
Because local newspapers published a relatively accurate list of the names
and addresses of most of the victims, it has been possible to overlay this
data on a 1909 street map of San Antonio and another that the US Geological Survey published showing the extent of flooding along the West
Side creeks and the San Antonio River. The resulting map offers a digital
expression of the contention that those inhabiting the city's most vulnerable and under-resourced section experienced the highest levels of death
and disarray, a skewing of burden and vulnerability that redlining in the
1930s would reinforce and extend well into the twenty-first century.[15]

Some of these short- and long-term concerns resurface in chapter 2, which explores the American Red Cross's remarkable efforts to provide essential supplies to those without any resources. Several days before the storm erupted over the Balcones Escarpment, the organization's Southwest Regional Office had been tracking its progress and preparing its staff for the possibility that they would need to decamp to San Antonio; key officials also forewarned the local chapter to begin gathering supplies and notifying volunteers. Their combined efforts entailed raising the funds to underwrite the distribution of food and clothing, construct and manage a tent city that was home for those without shelter for several weeks, and operate an employment center that helped the distressed find work. Although these and other commitments were universally praised and were absolutely critical at a time when the federal government had no mechanism to mount any such rescue operation (much as the Federal Emergency Management Agency does today), there was a less laudable backstory. Utilizing never-before-analyzed internal national Red Cross communications, the chapter reveals several tense debates and disagreements that complicated the relationship between and the actions of the national organization and the local chapter. Most troubling was the condescension if not outright racism that seemed to guide decisions about the amount, kind, and duration of relief the chapter—led by local elites—would sanction.

Even the US Army, which dispatched hundreds of soldiers into the city to pull people out of the river and creeks, and for several days afterward patrolled downtown streets and conducted cooperative search-and-recovery operations ("Sturdy boys, husky men, and Uncle Sam's soldiers with grappling hooks are stationed at the numerous bridges waiting for bodies to come down"), found itself caught up in the machinations of local politics.[16] Chapter 3 examines this complex, behind-the-scenes maneuvering. Because several city commissioners were savvier than Mayor O. B. Black, who had been elected to office just four months earlier, they used his inexperience to gain political advantage; the army was often the pawn in these internecine struggles. Extricating itself from local politics or fighting back—as one lieutenant did in an exposé of San Antonio's on-again, off-again relationship with those in military service whose paychecks filled local coffers—demonstrated the complex associations the flood revealed. Even from above: days after the floodwaters receded, the army sent pilots aloft to film the damage along the Alazán and San Pedro Creeks and the San Antonio River. This surveillance and the photographic mosaics it generated were central to the army's unpublished report on the flood, which offers dramatic documentation confirming *West Side Rising*'s arguments about the flood's brutal impact on the West Side.

Would the flood's brutality finally force San Antonio's politicians and its civic elite to do what no others of their station and standing had done after the floods that wracked the city over the preceding century? The quick answer is yes, and chapter 4 explores how the long history of ignoring the city's flood-prone nature finally gave way, and why the forces that had in the past deflected perceptions of the urgent need for flood protection failed this time to galvanize opposition. One catalyst for this shift was the sheer number of deaths that occurred. Another was the level of damage to the downtown core, and thus to the urban economy. It helped too that John Tobin, who in 1923 had replaced Black as mayor, was as skilled a political actor as his predecessor was not. Although Black had championed a local bond to pay for the construction of the Olmos Dam, he was unable to marshal the support to bring it before the voters. Tobin had no such problem. Elected in a landslide in May 1923, he won in part because he had campaigned relentlessly for citywide flood control, including robust intervention on the West Side creeks. That fall he and the city commissioners had agreed to a $4.35 million bond, with more than 60 percent devoted to flood prevention infrastructure. Tobin then trumpeted the bond's necessity in advance of the December election and cheered its relatively easy passage.

The completion of the Olmos Dam in late 1926 was a historic accomplishment, yet as is so often the case in human affairs, nothing is ever quite what it seems. That is why chapter 5 begins with a discussion of three warnings that questioned placing the city's faith in the dam—each of which was ignored. The first appeared in a *San Antonio Express* interview with W. W. Ashe, a visiting federal scientist and watershed specialist who had been trapped in the city by the 1921 deluge. He cautioned against

building a flood detention dam without also rigorously managing the upstream basin, noting that the overgrazed and heavily logged hills that were the catchment basin for Olmos Creek were responsible for immense amounts of silt and debris that ended up in the city's central core. As he -{ 17 }-
reminded the newspaper's readers, it was precisely this kind of material that a 1900 flood had flushed down the Colorado River, undercutting and overtopping the Austin dam, crashing into the state capitol, and killing eight people. Absent landscape-scale management, local dams were prone to failure. This was especially true if they were poorly constructed, a chilling point that consulting engineers raised with the city commissioners shortly after the Olmos Dam's dedication in December 1926. Their disturbing reports, now housed in the Water Resource Archive at the University of California, Riverside, apparently were buried.[17] Given what they detail, it is not surprising that these assessments never saw the light of day. Because the Olmos Dam was flawed in its design, built with a substandard quality of concrete and lacking a spillway and proper outflow gates, and susceptible to leaking at its base, the engineers were deeply worried that what happened in Austin was a real possibility in San Antonio. Although the Olmos Dam's precariousness would not have concerned investors who benefited from it by fashioning new elite suburban sanctuaries close to but above and behind it, had other venture capitalists known about the dam's structural problems, they might not have poured millions of dollars into the central core's many new tall buildings and modern skyline.

With a dam to its north and a straightened, deeper, and wider river running south, San Antonio's downtown boomed. The West Side did not. True, there had been modest investments in several of its creeks. The confluences of the Alazán and San Pedro, and the San Pedro and San Antonio River appear to have been restructured to carry greater volumes of water; some of the creek beds would be cleared of vegetation. None of these improvements, though, offered the kind of economic stimulus that elsewhere increased the property values and assets of the city's downtown merchants and landlords. The disparities intensified during the Great Depression and remained locked in place until the mid-1960s and early 1970s.

Chapter 6 details West Side residents' efforts across the second half of the twentieth century to alter their living conditions. The central actor was Communities Organized for Public Service (COPS), a powerful parish-based, female-dominated grassroots organization. COPS burst into local and national prominence following a relatively modest 1974 flood along Zarzamora Creek that turned streets into rivers and swamped houses. Tired of suffering from this form of slow violence and the political disregard that led to it, activists challenged the city to finally act on behalf of flood-weary neighborhoods. Their very public challenge developed into the organization's remarkably successful initiative that relatively quickly transformed the city's political structure, its budgetary commitments, and the health and resilience of the West Side community. Its activists later transplanted COPS's organizing methods and galvanizing message to Los Angeles, Houston, and other major US cities, making it one of the country's first political movements dedicated to environmental justice.

COPS's rising did not take place in isolation. The chapter also delineates earlier community-based organizations like Cruz Azul Mexicana, a female-led mutual aid society that rivaled the Red Cross in its relief efforts during the 1921 flood. Other influences on COPS included Representative Henry B. González, a child of the West Side who became the city's most powerful congressional representative to date by channeling millions of federal dollars into rebuilding the West Side creeks and funneling other monies into better housing, health care, and employment opportunities. His top-down efforts in the 1960s through the 1990s were reinforced by COPS's potent bottom-up energy, and rehabilitation of the people and the land proved a powerful incentive and a transformative force that helped resolve some of the key environmental inequities and social injustices that the 1921 flood in San Antonio had laid bare.

-[PROLOGUE]-

1819

Antonio María Martínez to Joaquín de Arredondo[1]

NO. 531

In the last week of June no special incident has occurred in this province under my command, and I notify you thereof for your superior information. God protect [you many years].

Béxar, July 4, 1819

NO. 532

The present situation of this province under my command is a lamentable one. This capital of Texas has undergone a ruinous disaster, and I am obligated to inform your superiority of the event which follows. Between six and seven o'clock on the morning of the 5th of the current month [July], a terrific downpour of water descended upon the head of this river causing a flood so terrible that no object could resist its fury. The river left its channel and spread so far over the town that it joined San Pedro Creek. The houses were caught between the conflicting currents, and the river crossed both plazas. After the stock in the *potrero* disappeared, the water began to dislodge and carry away several *jacales* from their locations, and the people who lived in them were seen as they were carried down the river by a strong current, and it was impossible to give immediate aid to the miserable souls who struggled against death because no one could do anything except look out for himself.

To give you a full account of the catastrophe would be a highly prolonged affair, but since you have sufficient knowledge of this town I will be content to tell you that from the proximity of the San Valero Mission to San Pedro Creek, which crosses behind the city on the West, it was all one river in a struggle with the houses and the stakes of the corrals. It formed an irresistible current, and in addition to the many houses it carried away, a good many people perished. They are being extracted from the river yet, and for this reason I cannot give the correct number of the total loss.

The force of the flood was so violent that no one could save his few possessions. Though both men and women had a great deal of agility in the water, each one was content with having saved his life. Almost all of them have nothing to clothe their bodies because the water carried away their house. Although they were dressed at the time of the floods, when they found themselves suddenly in the middle of the river they pulled off their clothes; some did so in order that their clothes would not interfere with their navigation. The result is that they are now in skins and have no food whatever to keep them alive.

These misfortunes were general, but the paymaster of the company of Béxar has been ruined because his granary is full of water and all the grain is fermenting. The continual rain will not permit him to move it to another place or even put it in the sun, and consequently much of it will be lost.

In this pitiful situation, finding myself unable to lighten such burdens, I solicit your mercy in the name of an unfortunate people who appeal to you as the father of these provinces, and I beg you, in order to alleviate their misery insofar as possible, to take the most effective measures that your meager resources will allow, understanding that whatever optimistic hopes the citizens and the troops had for next season's crops have vanished almost completely because the aforesaid flood has beat them down and submerged them in swamps. Therefore, it will be impossible for me to exist in this debris unless with charity and kindness you hasten to assist me and prevent the destruction of the troops and *vecinos* whose former needs I had attended by lending them corn from the warehouse so that they might do their planting, but now that this has all been lost they will hardly be able to pay me.

The bridge which facilitated communication with the company and the *vecinos* of the Alamo and which I had constructed last year with the ceaseless labor of those same *vecinos* and troops was the first thing carried away by the river, and consequently, if it continues to rain, those inhabitants will remain unable, as they are now, to communicate with this government. Since I am aware of your diverse needs and duties and the few resources on which you can depend, you must know that I shall appeal to the kindness of the Most Excellent Señor Viceroy, who, fearing further eventualities, will perhaps supply enough aid to compensate this province in part for the damages it has suffered.

The government building was badly damaged. The river passed by the side of it with indescribable violence, and it is now uninhabitable. It will be necessary to repair it, but I lack the means and resources, and I am greatly grieved to see myself living in a swamp.

All the troops are working devotedly to alleviate the condition of the people, especially the ad interim plaza attendant, D. Manuel La Fuente, *alférez* of the flying company of Alamo, and Manuel Menchaca and Juan Galván, sergeants of the company of Béxar. According to my order they administered to the needs of the town with the greatest energy while the flood was raging, and therefore they seem worthy of recommendation to your superiority. God protect you many years.

Béxar, July 8, 1819. By special mail.

The fatalities of the flood were, according to the Rev. Eugene Sugranes, CMF, in an account penned a century later, "six white grown persons, 'Gentes de Razon,' and ten Indian children." Sugranes recorded the adult victims—whom he labeled as either "maid," "widow," or "unmarried man"—as:

Concepción Davila	Maria Guadalupe Zambrano
Perfecta Hernandez	Francisco Montoya
Maria de la Cruz Hernandez	Josepha Vasquez

The children who perished were:

Juana Hernandez	Miguel Laro
Carmel Hernandez	Juana Laro
Juan Rodriquez	José Ruiz
Francisco Zambrano	Dolores Ramirez
Concepción Ruiz	Concepción Bastillon

Sugranes used the code "white" to describe the adults who perished, reflecting a practice in the 1920s for referring to Mexican Americans who wanted to "pass" for white.[2]

In the painful aftermath of the devastating September 1921 flood, still "laboring under the stress of the last eventful days when death, wrack and ruin seemed to reign supreme," the Rev. Eugene Sugranes wrote a lengthy article in the *San Antonio Express* challenging the belief that "the recent catastrophe that visited our fair city the night of September 9–10 was the most frightful and disastrous in the annals of San Antonio." He had read Gov. Antonio Martínez's heartrending description of the 1819 flood, and he urged readers to recognize the direct link between the two mega-inundations a century apart. "Let me state at the very outset, that the 1921 flood was neither the first nor the worst in San Antonio; but it ought to be, it must be, nay, it shall be the last." It would not do for history to be repeated again. That is why, "after a long and painstaking research through the dust-laden archives of the Old San Fernando Cathedral," Su-

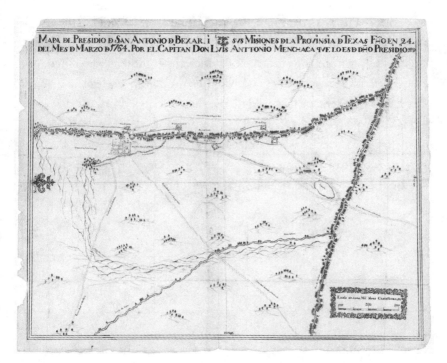

A 1764 plan of San Antonio shows the community located within the San Antonio River and San Pedro Creek watersheds. The map is oriented east–west, which is how the water runs, rather than the traditional north–south.

granes considered it "good fortune to find the record of the frightful" flood of 1819 and to be able to recover the memory of it.[3]

Despite Sugranes's need to dig into the archives to unearth these rarely—if ever—read documents, he insisted that the distant past had always been remembered, having been "handed down from generation to generation." Yet if such a memory of the earlier flood existed, it must have been pushed aside. This penchant to ignore the past, notes environmentalist Rob Nixon, dovetails with the "slow violence of delayed effects that structures some of our most consequential forgettings."[4] San Antonio's spatial layout embodies this studied disregard and its calamitous consequences. After all, the city in 1921 had been constructed inside the same floodplain as its 1819 predecessor, which in turn overlaid the initial early

eighteenth-century urban plan for the adobe frontier village tucked between the San Antonio River on its east and San Pedro Creek on its west,
flatland fully within the embrace of a sometimes violent watershed. Suc-
-{ 24 }- cessive generations, regardless of whether they lived under the Spanish,
Mexican, or American flag, added layer upon layer of *calles* and avenues,
plazas and squares, casas and cottages to the growing urbanizing environment that occasionally went underwater. Like flood-deposited silt, these
accretions built up, paving over the land and burying the once haunting
memory of what had happened in San Antonio in July 1819.

-{ ONE }-

"DEATH RIDES ON WATERS OF THREE STREAMS"

> General rainfall throughout all of Southwestern and Eastern Texas
> yesterday brought much needed relief from a prolonged drouth that
> had seriously retarded agricultural activity and was accompanied
> by a drop in temperature sufficient to bring noticeable relief from
> continued excessive heat. The total precipitation to San Antonio and
> vicinity amounted to .55 of an inch.
>
> —*San Antonio Express*, September 9, 1921

The rain came down in sheets. Then harder still, creating whiteout conditions. The thunder rattled windows and shook foundations; the constant booming "reverberated through the heavens." The sky seemed to be on fire, with an "electric display [that] was the most vivid ever seen" in San Antonio.[1] Whether people huddled in tiny shacks in the low-lying neighborhoods on the West Side, bedded down for a long night in comfortable cottages near San Pedro Creek, or drew the curtains of large homes on higher ground above the San Antonio River, they could hear the rain. Later, they would learn the cacophony was the result of a tropical depression that had stalled over the region, dropping more than ten inches on the upper reaches of the Alazán, Apache, and Martínez Creeks; seven inches on the narrow watershed of San Pedro Creek; and an astonishing seventeen inches on the hills ringing the Olmos Basin, whose tributaries formed Olmos Creek, which fed into the headwaters of the San Antonio River. "While torrents of rain were still falling in the streets of San

Antonio," a breathless *San Antonio Light* reporter recounted the next day, "and the residents, unable to get out because of the downpour went early to bed, a roar was heard, subdued but ominous as the floodwaters broke upon the town."[2]

The stormwaters hit the city from two directions, causing two vastly different outcomes. The deadliest flooding came from the narrow West Side creeks flowing out of the low hills to San Antonio's northwest. At the same time, the San Antonio River, whose watershed is defined by the Olmos Basin north of the downtown core, was breaking over its banks. The disparate contexts and consequences of these parallel events mean there are two narratives to the 1921 flood, one of which has taken precedence up to now, but which told together reveal a deeper and more complex story.

The Zepeda family never stood a chance, though it was the hope of increasing his family's odds of survival that led Juan Zepeda to act the moment he sensed water flowing inside their cottage. The water's presence indoors meant that after a night of torrential rain, the nearby Alazán Creek had poured over its narrow banks.

The Zepedas' neighborhood was surrounded by a dense cluster of shedlike, one-story, thin-framed homes, often housing as many as three families under one roof. An equally large number of people in the neighborhood crowded into "corrals," an infamous form of unsanitary housing typical of San Antonio's West Side. Once stalls for horses and mules, the structures had been converted by slumlord owners into human domiciles that were dank, dusty, and dark. The large and impoverished barrio tucked in close to the South Laredo Street bridge, which was among the first to feel the punishing impact of the September 1921 flood. "This territory was in the direct path of the oncoming rush of flood water," the *Express* later noted, and "the district was swept entirely out of existence, with the greatest loss of life and property occurring along the Alazán."[3]

Confirmation of the devastation came the next day when a reporter for *La Prensa* trekked west along Mitchell Street near where its wooden bridge crossed San Pedro Creek. The corner store in the one hundred block at the intersection with South Flores Street "was blown up by the

Some of the wreckage that Alazán Creek piled up against the International & Great Northern trestle.

San Pedro stream's current and 20 yards from there the bridge collapsed to pieces, leaving only half of the bridge still standing. The other sections were completely lost and knowing that some people died there, later on the police picked up some cadavers of unrecognized people." One block farther, the damage seemed to intensify: "Some of the houses of Mexican families who live there and others located toward the back disappeared from that same stream without any vestiges remaining." At the same spot, the creek dumped "a good amount of cotton bales from the Alamo Ginning Company, leaving some trapped in between corpulent trees and the rest of the wooden houses that went under." There too the current "threw 'Ford' automobiles to the shore as well as many dead horses. This location only offered aspects of desolation and sadness."[4]

Elsewhere the news was just as grim. A journalist for the *Express* walked the full length of Alazán Creek and recorded its wrecking-ball-like power. Its waters pushed houses and stores off their foundations and cleared away full blocks, shoving structures into nearby buildings. In their wake, the floodwaters left behind impassable streets and alleys jammed with a "huge mass of timber, roofs, driftwood and other debris." The littered creek bed was filled with the corpses of "cats, goats, chickens, guinea pigs and other pets." The farther southeast the reporter went, the greater the damage. Dozens of houses had piled up against the International–Great Northern bridge, forcing floodwaters to race down adjacent streets and accelerating the spread of devastation. Those living south of the bridge were not spared: a quick count of the houses wiped out where the Alazán crosses South Laredo Street tallied more than sixty. And where the creek emptied into the San Pedro near the Union Stock Yards, a vast churning lake thought to be five hundred yards wide killed more people and animals. The reporter bore witness to what amounted to a two-mile swath of destruction, a "clean sweep."[5]

The results of these two fact-finding missions would be updated. Later, if still tentative, estimates of the number of homes destroyed along the Martínez, Apache, San Pedro, and most especially, the Alazán rose to 1,500. At a minimum, the number of people that police and fire officers, volunteers, and soldiers pulled out of the swirling creek waters during the chaotic night and predawn hours was more than 2,000. The unimaginable damage received vivid aerial depiction when US Army pilots flew over the Alazán watershed and photographed how easily the ferocious floodwaters had snapped homes into kindling and left "hundreds of Mexicans who live in this vicinity in small shanties...homeless."[6]

These homeless, who escaped the torrent, were the lucky ones. Others, "caught unaware, sleeping," were crushed by debris or drowned.[7] How Juan Zepeda recognized, in the dead of night, that his South Laredo Street home was in great danger is impossible to know. But as survivors later told reporters, anyone who lived close to the West Side creeks was used to floods. Every time it rained, their roads turned into thick goo, and if the deluges were heavy enough, creeks leaped out of their banks, running through the thickly settled barrio. In October 1913 four people died; three months later, another powerful flood killed nine. Yet no one recalled the Alazán rising as swiftly as it had late that Friday night; no one remembered it producing such stunning levels of havoc. Still, something had warned Juan Zepeda. Perhaps it was the noise of pounding rain and distant cracking of wood or perhaps a scream. Whatever alerted him, he woke his family and in-laws and sloshed outside to harness a horse to their wagon. He hustled his father-in-law, wife, and seven children (one of whom may have been a cousin) through the blinding rain to the wagon, water swirling around its mud-encased wheels. With everyone safely aboard, he attempted to cross the Alazán. That is when a house, racing downstream on a wall of swift-moving water, slammed into the Zepedas' wagon, flipping it over. Its ten passengers were sucked beneath the churning waves; only the father-in-law escaped, along with the horse, which had managed to shake loose from its harness and find its footing.[8]

The Zepedas died late Friday night, but by Monday morning only seven of their bodies had been recovered for burial that day. "Perhaps the saddest funeral was that of the Zepeda family," the *Express* noted. "A mother and father and five little children ranging in age from seven months to twelve years were laid to rest in the same grave, and there is almost a certainty that two more of the children will be brought to the family grave when they have been found, as they have been numbered among the missing for two days." When the missing children were located and identified, they were buried with their family. One of the mourners, Tom Garcia, Juan Zepeda's cousin, told a reporter that the family's home "was still standing Saturday morning after the floodwaters had subsided."[9]

The Zepedas' burial was among a record number of funerals September 12 at San Fernando Cemetery. Three members of the Hernandez family who had lived on South Flores Street—mother Elena (age twenty-five), daughter Estella (six), and son Adolfo Jr. (four)—were also interred together. A husband and father, Juan de la Garza, may have made the same choice for his family's interment—his wife, Petra (thirty-five), and their children Onesema (four), Carlota (three), and Lontardo (one), had been pulled into the roaring San Pedro from their creek-side abode

-[30]-

ALL THAT REMAINS
OF BUENA VISTA
PLAY GROUNDS
ALAZAN CREEK

Alazán Creek destroyed the Buena Vista playground.

on McAskill Street. They were the first among many who would be buried at the cemetery, where on that Monday "prayers were said...for the repose of the souls of 25 flood victims." *La Prensa*'s report continued: "From the early hours of the morning until the shadows lengthened in the late afternoon, it was an almost continuous processing of relatives who came to bury their dead. In the already thickly populated silent city freshly marked mounds and newly dug graves told the story of the frightful death toll of Friday night and Saturday morning." At each grave site, the bereaved heaped flowers: "No matter how humble the victims there were floral offerings in profusion, and in a little while there will arise new black or white crosses to mark their resting place."[10]

Not all the victims were so memorialized. There would be no markers of any kind for those whose bodies, battered and pulled from the debris, could not be identified. The brief descriptions printed in the daily "Death List," compiled from local mortuaries' records, serve as haunting, telegraphic eulogies:[11]

"Woman found near Union Stock Yards."

"Mexican Woman about 35 years old."

"A white man about 40 years of age, found near the fairgrounds. It was wedged between the forks of a tree in the main channel of the San Antonio River. The body was swelled from exposure from the sun, and three men had difficulty dislodging it. Because of its decomposition, it was buried where it was found immediately. Identification was impossible."

"Unidentified Mexican boy. Near the Union Stock Yards."

The missing simply vanished. They received no last rites. No crosses. No flowers. They did not disappear from their frantic families' memories, though, and it was not for lack of trying that their remains were never found. For days, grieving relatives searched for their loved ones along the fetid, junk-strewn banks of the creeks, picking through the wreckage there and along the lower reaches of the San Antonio River. The names of the missing appeared in the local papers. Among them was two-year-old Teodora Cardenas, who lived with her family at 540 Mitchell Street; only the body of her sister, five-year-old Tura, was located. Then there was Caravella—no first name or gender indicated, age forty, who resided at 1118 Monterey Street. Although Gertrude Southall had a first and last name as well as a known address—1619 Lakeview Avenue—her only other identifying feature, according to the newspapers, was "negress." (Her body was later recovered from under debris that piled up against the International–Great Northern Railroad trestle crossing Alazán Creek.)[12]

In the case of other victims, it was as if they had never existed at all. This was the case most chillingly for two families who had lived south of the Mitchell Street bridge, which had spanned San Pedro Creek before the violent floodwaters splintered it. The surrounding area had been doubly inundated, as the creek's roiling waters boiled into and backed up with those of the San Antonio River. Their combined whirlpool-like force pulled down homes and buildings, leaving behind immense devastation. At some point Friday night, the Morales family vanished. Only the body of Francisca, age thirty-six, was recovered; she was found more than three miles downstream, close to Mission San José. Neither her unnamed husband nor her seven children were located. The fates of Tom McCray, his unnamed wife, and their six children—two girls and four boys—were as obscure. "Relatives unable to find any trace of the house or anything pertaining to it," read the telegraphic-like accounting in the Death List. "Team of horses and wagon also missing."[13]

Like Juan Zepeda, many of the flood's victims took action to try to survive the onslaught—the floodwaters bore evidence of these attempts. "Pitiful and mute witnesses of frantic efforts of people to save themselves" from the West Side creeks' ravaging energy were "bits of drift that came hurtling down the streams last night," the *Express* reported. "Caught on one bridge was a rude raft of old white enameled doors bound together. Tied securely on one of these was the battered and wretched sleeve of a child's coat. It had been lashed on with the cord of an electric iron, which evidently had been hacked with a kitchen knife." Another flotation device, consisting of "a collection of planks tied together in which were fastened several automobile inner tubes," ended up on a street corner. The rubbery material was "torn into ribbons," a stand-in image for the probable fate of those who had clung to the makeshift vessel.[14]

Grasping at whatever they could, some people rode out the high waters. Mrs. Ralph Saenz never made it out of her home at 121 Cass Street "near the fatal South Flores Street crossing." When the Alazán's high waters smacked into her house, she immediately grabbed her "little niece" and lifted her into a small attic, a refuge Saenz could not reach as the waters rapidly filled the room: Saenz clung to the top of a door for nearly two hours.[15] Those torn from their flooded shacks found themselves being "hurled along by the current until they landed where their feet again touched the ground" and then scrambled to find a drier patch of earth. Still others snatched at trees or telephone poles, hoping to claw their way up above the furious, debris-filled waters. It was an instinct imprinted on history: nearly every previous flood that ripped through San Antonio featured tales of people who had saved themselves by climbing a tree.

That form of sanctuary proved especially critical along South Flores Street, just south of the confluence of the Alazán and Apache Creeks, which was hit with an eight-foot surge of water. Its sudden onset gave residents only moments to react. For dozens, trees offered their only respite. A crowd of onlookers "witnessed the frantic efforts of a man who dimly could be seen by aid of police search lights as he clung to limbs of trees in an effort to keep himself from the rush of the torrent." Yet as he tried to climb above the water, he slipped from view. "Those nearest the point where the man was seen, state that a child or woman was across his back, making his struggle all-the-more difficult. Little hope is held that he escaped to safety."[16]

No one would have been surprised if young Francisco Gutiérrez had also died while fighting to live. He, too, resided on South Flores. Like many of his neighbors, the twelve-year-old was pulled into "the current that rushed at a speed that destroyed all in its path." Flushed from his family's collapsing house, he heard his five-year-old nephew cry out and managed to find him. Amid the roaring waters, Gutiérrez spotted a small tree, and as the pair swept past it he caught hold of the trunk; with his nephew on his back, he clambered up as high as he could. For five hours or more, he held on tight even though he was "battered black and blue by floating wreckage." When the floodwaters finally receded and the boys were rescued, the "nearly unconscious" Gutiérrez was taken to Robert B. Green Memorial Hospital to have his injuries treated. The *San Antonio Express* called his dead-of-night initiative "perhaps the outstanding heroism of the flood."[17]

Others proved as fearless. Intermingling with units of the fire and police departments were soldiers and West Side residents, men and women who waded into the racing creeks to pull people out of harm's way. Alfredo Gutiérrez, stationed at Camp Normoyle, is credited with rescuing dozens from the confluence of the Alazán and San Pedro Creeks.[18] Members of West Side mutual aid societies—Cruz Azul and Sociedad Hidalgo, dedicated to providing health care and other social services to their community—performed valiantly; Cruz Azul was especially active, providing clothing, medical aid, and solace until late November.[19] Pitching in, too, were those who by happenstance were caught up in the same floodwaters as the people they helped rescue. Struggling against the turbulence in a desperate search for his wife, Jesús Cardenas came upon a weakened Anita Gutiérrez, Francisco's mother, and her married daughter, Esther Draeger, herself a mother. "Cardenas helped them climb up his shoulders and, in this way, form a ladder. They managed to climb up to the high branches where they waited for other help." Mother and daughter were rescued near dawn, but Draeger's nine-month-old daughter, Alice, had been ripped from her mother's tight grasp.[20] Porfiria Acosta, who owned one of the small houses near the Alazán, a home that "represented the entire estate of this woman," nearly lost her life when the structure was "dragged by the flood currents and completely disappeared." If not for the "timely help" of Leonardo Flores and Agustin Garcia, she would have died.[21]

The loss of lives and livelihoods, of hearths and homes, hit the West Side hard. More than eighty people died or disappeared that night. This total surpasses the fifty-one deaths due to drowning that the city reported in its final accounting in October.[22] The disparity lies in the manner of reporting. The city did not include those unidentified or missing in its final tabulation, and this undercount, whether deliberate or otherwise, enabled city leaders to assert that San Antonio had not suffered a catastrophic loss (and subsequently to mount a public relations campaign to counter media reports that suggested the city had been destroyed).[23] The count also did not provide a breakdown of who died and where, though it was easy enough to decipher. Approximately 95 percent of the fatalities had lived in the neighborhoods the West Side creeks had gutted. Of these, more than 85 percent had Spanish surnames (with others of mixed parentage); women and girls died at a higher rate (60 percent) than men and boys.

Even the smallest percentage contains a larger story. Although only one African American was identified in these statistics—Gertrude Southall—the flood battered the Black community on the West Side. Many of its original members had migrated to the district during the post–Civil War era, replacing members of the white working class who left the neighborhood in an early version of white flight. The emerging community, which was home to several churches, was dubbed Newcombville, perhaps honoring or deriding James Pearson Newcomb (1837–1907), a Radical Republican who for a time was a staunch defender of Black civil rights in San Antonio. By 1920, according to the US Census, upward of four thousand African Americans lived in an area loosely defined by Commerce Street on the south, Lombrano Street on the north, San Pedro Creek on the east, and Alazán Creek on the west.[24] This siting proved troublesome on the evening of September 10. "The first washout on the Alazán Creek, beginning at West End Lake, was at Castro Street. A negro church and a four-room house were washed away here. The church was carried two blocks down the creek, where it was lodged."[25] As the Alazán's floodwaters coursed in a southeasterly direction, they crossed Zarzamora Street and swept away another "small negro home" on Calaveras Street and two more on Rivas and Delgado Streets. The greatest destruction occurred between Arbor Place and West Travis Street, where "numerous negro and Mexican homes" were destroyed. "Four houses on the corner of Perez and Salinas Streets and six more were lost where Lakeview crosses the creek." This dangerous intersection of Lakeview Avenue and the Alazán is where Gertrude Southall had lived and died.[26]

Her demise, as with that of all the others on the West Side, was given human form in a *New York Times* front-page story, under the subhead "Mexican Quarter Covered": "Babies were swept from mothers' arms and lost, mothers were carried away and children rescued. Fathers were lost saving little ones, and today there are widows and orphans in San Antonio who shudder at the thought of last night."[27]

It is no wonder, then, that survivors bore all the signs of emotional trauma. A child lashed out at those who had snatched him from the riotous waters because they had not been able to save his younger brother. The grief of grandparents, parents, and siblings was so intense that it left them mute, insensible to expressions of sympathy. Physically battered and emotionally scarred, they stood in silence, in long lines waiting for a hot meal, a warm blanket, or a set of clothes. In one photograph, a child holds a younger child on her hip. Straight into the camera's lens, her stare speaks volumes.[28]

English-speaking journalists—less than two years after the Great War had ended, when the concept of shell shock had entered common parlance—conflated these bodily responses and affective reactions with racial characteristics. One reporter linked people's impassiveness to their

Children gathered at a US Army food wagon in the aftermath of the 1921 flood.

ethnicity. Another believed that "race" explained why "some of these homeless Mexicans" were able to bed down wherever they were; they "merely used their burden for a pillow and then fell asleep." A third noted the survivors' instinctive generosity: "There was this about the Mexicans of either sex—loyalty. No matter how small the pittance they might have grabbed in dashing out of the dangerous waters, they were willing to share it with those less fortunate fellow sufferers." Traumatized, "these folks are all pauperized. They will need all of the help more fortunate San Antonians, the Red Cross and charitable organizations can give them. And they will be grateful."[29]

These stereotypes defined the part of the city known variously as the Mexican settlement, district, or quarter—terms that segregated and divided. The gruesome cleanup reinforced the perspective that the West Side was a place apart. To avoid the spread of infectious diseases from decomposing animals, sanitation workers had the thankless task of collecting chickens, dogs, rats, and other small animals and burning them; large mammals—horses and livestock—were incinerated where they were found. For several days a funeral-pyre-like pall of rank smoke hovered over the West Side. Then the crickets descended. Freed from the competition for food, these rapidly adaptive insects became a noisy plague across San Antonio and other flooded regions of the state, scavenging in biblical proportions. What should have been a unilaterally uplifting sight—a nonstop migration of billions of snout butterflies, often seen filling the city's skies after heavy rains—was instead considered overwhelming, an invasion, and interpreted as an omen of a hard winter ahead.[30] The West Side had become a cautionary tale, as was evident in the accounts that filled the newspapers. One journalist, spotting a garish, "immense red pillow" precariously perched on top of the twisted International–Great Northern trestle over Alazán Creek, wrote: "A red flag, a red undershirt, has ever been a signal to travel cautiously, to heed the warning and stop."[31]

To leave the story there, with a neatly wrapped metaphor that suggests a traumatized people rendered speechless and dependent, whose immediate needs could be met only by the succor handed out by the well-to-do, would be to miss their voices and agency. This disaster, after all, was not simply a consequence of a rogue tropical depression. Its corrosive floodwaters exposed the strikingly different experiences felt by the city's haves and have-nots. They revealed, too, the long history of such inequity. Some of these differences were written into the spatial inequalities that flowed along lines of class, ethnicity, and race and that determined the odds of who lived and who died during the flood. That was a key argument in "La tragedia de la inundación de San Antonio," a powerful, interview-based, sixty-four-page illustrated report published within weeks of the flood. It provides the best single-source documentation of the flood's catastrophic impact on the West Side community.[32]

The social and economic disparities that defined San Antonio, the report's anonymous authors argue, were made explicit in the different kinds of damage the river and creeks produced: The "San Antonio River hit the rich—it affected the big stores on Avenida C. The powerful houses of Houston and Commerce St. It must be said in its honor that it was

LAS INUNDACIONES EN SAN ANTONIO

Las que la Ciudad ha sufrido en una Centuria.

"Las inundaciones en San Antonio," which ran in the September 18, 1921, issue of *La Prensa*.

greedy—it wanted riches and destroyed estates."[33] By contrast, Alazán Creek—"an imitation of a brook, a laughable pantomime, a thin and flexible snake"—proved ravenous. "It was the taker of lives—it was a cruel executioner who wiped out every poor soul it encountered." Put differently, the river

> swallowed pianos, velvet rugs, Venetian moons of unparalleled beauty and wealth. Alazán Creek drowned children, killed women, knocked down men. And it was our people, the Mexican people, that succumbed defeated, whose poverty did not allow [them] to reside in a house in a pious neighborhood, a street near the center and out of danger. The sons of Mexico were the ones that fell asleep, unperturbed by danger, to wake up in the hands of a monster.[34]

The details in this strikingly blunt analysis were predictive of the story of the flood as mediated by local English-language newspapers within days of the disaster. Reporting first focused on the devastation wrought by the West Side creeks, yielding a sympathetic narrative, but that began to change with the falling water levels of the streams. By the third day post-flood, local media were constructing a competing account that prioritized the San Antonio River's ravaging of capital and commerce in the business district. This served a new purpose: to prevent another flood from destroying the urban marketplace, the newspapers supported those advocating the spending of a vast amount of public money to rebuild, protect, and sustain private entrepreneurial interests and their economic advantages. In this, San Antonio's political and commercial elite acted much like their peers in the other flood-prone cities of New Orleans and Sacramento, choosing to preserve their economic interests first and foremost.[35]

To justify this self-protection and the skewed economic outcomes that would flow from it, the narrators of this discourse—San Antonio's Anglo leadership—championed the inestimable value of the urban economy and its wealth-creating productivity. Although the poor's rent and labor contributed significantly to the local economy's success, they did not figure in this emerging narrative about the city's rich future. To achieve this glittering promise required surmounting the obstacle posed by a rampaging San Antonio River. After all, like the West Side creeks, the San Antonio had packed a punch. Those inhabiting the late-nineteenth-century streetcar suburb of Alamo Heights, "sitting like a watchtower, with its handsome residences and exclusive air," recalled the river's thunderous sound. Residents of the neighborhood, which lay on higher ground, awoke to a fear-inspiring roar, what they thought was a cyclone. "They found no wind but saw a dull colored avalanche sweeping from the Olmos and the San Antonio River into the valley which houses San Antonio."[36]

Uncorked, the silt-thick, debris-heavy floodwaters exploded through Brackenridge and Koehler Parks around 10:30 Friday evening, chasing upward of a hundred campers from the oak-studded grounds. Some of the flow broke east, where it joined with water streaming down from Fort Sam Houston and Government Hill and pushed a two- to three-foot surge down River Avenue (now Broadway). As the floodwaters reclaimed the San Antonio's historic floodplain—"River Avenue Lives up to Its Name," one headline blared—the water inundated street-fronting residences, apartment buildings, and commercial establishments; swamped the river-hugging Pearl Brewery; and pushed into the eastern quadrant of the central core framed by East Travis and Houston Streets, Avenue C, and Crockett Street. Meanwhile, the main thrust of the flood had plowed straight downriver, though given the San Antonio's serpentine course, "straight" is a misnomer. Wherever stormwater hit a sharp bend it overtook the bank, following its own direct line. By midnight, four to six feet of water was churning down Saint Mary's Street and other north–south conduits before pooling on lower ground. With trees and cars, furniture, pianos, and cribs roiling the waters, they acted like a battering ram, knocking out bridges, punching holes in buildings, shattering windows, and toppling oil tanks. As the water gouged the streets, it forced up hundreds of heavy wooden blocks that had been laid down to harden many of the city's major thoroughfares, compounding the maelstrom's destructive capacity. By 3 a.m. Saturday, the worst seemed over; having hit its crest, the flood, which in some places measured more than twelve feet deep, slowly began to drain away.[37]

Houston Street floodwaters on September 10, 1921.

In their retreat, the river's floodwaters revealed the damage left in their wake. Four residents had succumbed to the river's terrifying rush. The small number of fatalities in comparison to those in the West Side did not diminish these families' grief. The Frausto children, with their parents, an aunt, and five other children, had climbed nearby trees when the river barreled down Jones Avenue, inundating their residence. At some point two of the children, Ramón and Hortencia, fell into the water. The next day a soldier found Hortencia, and she was buried that Sunday. The bereft parents were startled to learn, after returning from their daughter's funeral, that a neighbor had discovered Ramón's body beneath a mound of wreckage several blocks from their home. As for West Side resident and fruit vendor Ben Corbo, the San Antonio's floodwaters roaring down Saint Mary's Street had trapped him beneath the rubble.[38]

Even as these victims were laid to rest, the main storyline of the flood centered on the San Antonio River's disruption to business activity. That was the overriding subject of a special meeting of the City Commission on September 11 and of newspaper accounts about the damaged central core. Basements were flooded, making it impossible to calculate the cost

Debris jammed against the Saint Mary's Street bridge near where Ben Corbo died.

of the flood to banks, department stores, and other consumer services; once "inventories of stocks are taken and the value of salvaged goods is estimated," the *Express* expected the tab to run in the hundreds of thousands, if not millions. The Saint Anthony and Gunter Hotels, which had taken on water like sinking ships, were a mess, as were the major theaters—the Empire, Princess, and Royal. The commodious Methodist Tabernacle was gutted; both side walls were "torn away, and all fixtures in the building floated out with the receding flood." Just blocks away, Saint Mary's Catholic Church was also unsalvageable. Its foundation was made of limestone, and the flood had weakened its structural integrity; the church subsequently would be torn down. Smaller buildings made of adobe, once the city's main construction material, would also be dismantled. Dozens of these buildings, "ranging in age from 50 years to nearly a century," had collapsed and had to "be cleaned from side streets." (If the age of the oldest of these adobe buildings is accurate, then they would have been constructed shortly after the 1819 flood.) Power outages had shuttered restaurants, melted ice, and stopped elevators, making for "weary flights of

This automobile, one of hundreds swept away by floodwaters, came to rest at the parsonage of Saint Mary's Catholic Church.

stairs to be climbed in downtown office buildings that do not have their own power plants." Streetcars were stalled, and telephone and telegraph communications were hit-or-miss, complicating a speedy return to work-day routines. So many railroad trestles had been damaged that passenger and freight service was suspended. Further complicating travel in a city of countless bridges was that nearly every one of the viaducts had been dam-aged, many beyond easy repair. Not that car-owning San Antonians could drive down most of the central avenues and streets anyway. The flood had dumped so much material that reporters had a field day picking through the flotsam and jetsam to record their oddities. "Here and there where driftwood is piled in some places as high as twenty feet, are beer kegs, babies' high chairs, Victrolas, butchers' meat blocks, bed steads, all a con-glomerate mass of wreckage, telling mutely that the elements, when loosed from the gates of hell, are no respecters of persons, class, or creed."[39]

That claim was inaccurate: the flood could not have been more re-spectful of the well-heeled. They may have lost things, even a good many of them, but their families and friends did not vanish beneath turbulent waters; their neighborhoods were not erased. The imbalance of pain and suffering found additional expression in the vast disparity in property losses between the central core and the West Side. Downtown, Houston Street was particularly hard hit. The owner of S. Rabe Antiques, in the Gunter Hotel, who boasted that his enterprise was one of the largest in the United States, offered a "conservative estimate" of $125,000 in losses. General manager J. J. Sterne of the Wolff & Marx department store gave a rough calculation of $90,000 in damaged stock housed in its completely flooded basement. Employees had not yet inventoried "the stock on the first floor, where the water reached as high as the second shelves of merchandise and submerged the show cases on the floor." The iconic Manhattan Café, whose bakery, storerooms, and stockrooms were located in the basement, suffered a total loss of its inventory, roughly $60,000.[40]

This red ink, individually crucial to each manager, operator, hotelier, and shopkeeper for insurance purposes, was important in two other respects. No bookkeepers tabulating the losses for West Side businesses came close to racking up the same high-value inventories as their downtown counterparts. No one was confused, either, about which sector was the subject of headlines in early broadsheet bulletins published while the floodwaters had yet to recede: PROPERTY LOSS IMMENSE; BUSINESS LOSSES EXTRA HEAVY; BUSINESS SECTION MASS OF WRECKAGE AND STOCKS RUINED.[41]

Moreover, when the numbers from the swamped business district were combined with the costs the city and property owners incurred for general cleanup, first responders' overtime, street resurfacing, and utility repairs, among innumerable other immediate reconstruction needs, Mayor O. B. Black's early guess that property losses amounted to $3 million seems low. Whatever the subsequent total turned out to be—and some estimates soared as high as $7 million—the figure fueled the demand that San Antonio finally build a retention dam to control Olmos Creek. The business district was simply too valuable to remain vulnerable to the whims of nature. That was the implication of Mayor Black's declaration on the first day post-flood that the city must float a bond for as much as $5 million "to relieve the present situation in San Antonio resulting from

the inundation of Friday night, and to build a dam across Olmos Creek."
Only in this way would the city's economy recover; only then would outside investors and local entrepreneurs feel confident in the city's commitment to protecting property values and capital formation. "All danger of
another such overflow as the one just gone through must be removed."[42]

In support of the mayor's belief that the dam would have tangible
economic consequences, C. H. Kearney, a former city commissioner and
"one of the best known engineers in the city," argued that the return on
the investment in its construction would be immediate: the "cost of such
a dam would have been small in comparison to the loss resulting from the
flood of waters sweeping over the city Friday night."[43] A dam, in effect,
would pay for itself, supporters argued—a claim that found quick favor
with the powerful chamber of commerce, the flood prevention committee,
and other bodies whose members included the city's business and commercial elite. As it organized the collection of funds to purchase food and
supplies, and, curiously, "handled the task of checking and verifying all
reports of dead, injured, and missing," the chamber met repeatedly with
the mayor and city commissioners to advocate for the dam. The chamber
did not need to lobby local print media, whose publishers belonged to the
organization; their inundated pressrooms activated their keen interest in
making the dam a reality. "The community must not allow its rivers and
creeks to run wild any longer," the *Express* asserted. "Whatever the means
of providing the protective engineering work in the Olmos, bond issue
or otherwise, San Antonio must now take heed of its death list and its
property destruction—and build for its vital protection."[44]

The city's power brokers understood that a dam's protection extended
only so far—it would not shield the West Side. For those creeks, they
drew on recommendations from a seminal 1920 engineering report the
city had commissioned from Boston firm Metcalf and Eddy, and which
the *Express* dutifully highlighted one week after the flood. The report
noted that the West Side's topography would not support damming
the creeks; that that sector's negligible land values, by themselves and in
comparison to the business district, did not warrant investing in significant flood control infrastructure; and that it was most cost effective to

cut down trees and other vegetation growing in the creek beds. Over the next two decades, according to the report, as funding allowed, the city should deepen those creeks' channels and widen the confluence of Alazán and Apache Creeks, their merger with San Pedro Creek, and where their combined flow poured into the San Antonio River. The 1921 flood may have ramped up the urgency for reconstructing these creeks, if only to avoid yet another painful "Death List," but by publishing the report's conclusions the newspaper effectively prioritized the dam's construction on the main channel of the San Antonio River and endorsed the go-slow approach to rehabbing the creeks.[45]

What the downtown boosters did not acknowledge was that had they also invested significantly in flood control on the West Side, this would have enhanced that sector's economic activity and value—which is what they expected the dam would achieve for their downtown resources, investments, and assets. But the Anglo civic elite had no expectation that the conditions on the West Side would be substantively changed. There had been one proposal urging the city to take advantage of the destruction of the corrals and shacks to replace them with better housing. J. B. Gwin, a Red Cross professional who managed the organization's regional response to the disaster, offered this solution in an article in the *Survey*, but his conception of a revitalized West Side gained no hearing in city hall or at the chamber of commerce; even the head of the local Red Cross chapter dismissed Gwin's idea as well-meaning but unnecessary.[46]

Instead, local media valorized two individuals whose life stories seemed to be critiques of the beleaguered inhabitants of the West Side. The first narrative had all the markings of a morality play, in this case involving a mother, a son, and a duck. F. J. Reimschissel, "a painter and paperhanger by trade," lived with his paralyzed mother, Augusta Morris, in a cottage on Alvarez Street close to Alazán Creek. The floodwaters filled the house and lifted the building off its foundation. As the structure began to drift with the torrent, Reimschissel punched a hole in the ceiling and managed to lift his disabled mother to the roof, which was barely a foot above the water's surface. Carried to Cass Street, the submerged cottage bumped along a fence, finding an unsteady anchorage. That is when a duck joined them, waddling onto the water-sloshed roof and remaining "with the two human beings, who fought off death" until they were rescued around 4 a.m. Saturday. Like so many of their neighbors, Reimschissel and his mother lost everything. But unlike so many others, Reimschissel did not go to relief stations for food, clothing, and shelter; "exemplifying the spirit of stolid Americanism, he refused Red Cross or other aid."[47]

One could read between the lines: "real" Americans—a code for whiteness—were self-reliant; charity bred dependence. The set of traits linked to utter self-reliance, long championed by social Darwinists, was particularly associated with the kind of person William Graham Sumner championed as the "Forgotten Man." "He is the simple, honest laborer, ready to earn his living by productive work. We pass him by because he is independent, self-supporting, and asks no favors," Sumner wrote. "We do not remember him because he makes no clamor; but I appeal to you whether he is not the man who ought to be remembered first of all, and whether, on any sound social theory, we ought not to protect him against the burdens of the goodfornothing." In newspaper accounts, Reimschissel seems perfectly cast as Sumner's imagined hero, but the celebration of his compete self-sufficiency also offers an insidious critique of presumed West Side dependency. His putative virtue, and the "Americanism" he embodied, subtly emphasized just how "non-American" his neighbors were.[48]

One reason Mexican Americans were unable to alter their second-class status emerges in another newspaper account that reads like a parable on the virtue of the self-made man. Channeling Horatio Alger, the *San Antonio Light* recounted the uplifting story of a local entrepreneur, W. E. Knight: "When but a tiny lad, Ed. Knight struck out alone in the world to shift for himself. Nerve, love of work, active and possessed of that ability to save rewarded his years of struggle in life's battle. He won." His victory was evident in his success in home ownership. His "beautiful home" on City Street in the King William district stood as "a testimonial of what a penniless lad may accomplish by will power." He could afford this sumptuous manse in part because he owned twenty-five "small rent homes" whose tenants' monthly payments "provided an income that assured him days of rest in his advanced age." What he owned, in fact, were shacks on the West Side, making him one of the class of exploitative landlords in San Antonio who profited from and deepened the barrio's immiseration. That is not how the *San Antonio Light* described the situation, though. Rather, in the aftermath of the flood, it commiserated with Knight's loss of an invaluable income stream: "The flood waters within a few hours deprived him of these possessions," and his splintered shacks "now form a small bit of the mammoth pile of wreckage in the vicinity of the I. & G.N. railway trestles." Even though he had lost what he estimated was a $27,000 investment, Knight vowed to bounce back: "You can't keep a good man down, if he won't stay down—I am able to start over again."[49]

Knight had access to human capital and financial resources that his tenants could never tap, resources that enabled him to rebuild a series of small shacks that looked much like those the flood had destroyed. His peers, who resided in tony King William or leafy upland suburbs such as Monte Vista, Tobin Hill, and Alamo Heights, also reconstructed what they had lost. Within a relatively short period, the West Side housing stock looked as it had on September 8, 1921. By the 1930s, the appalling conditions of the area—still flood-wracked and suffering from dilapidated housing, high levels of communicable diseases, no running water, sewage, and unpaved streets—was attracting shocked national attention. One journalist lambasted San Antonio in 1939 as "the shame of Texas."[50] The "slow violence" that had enveloped West Side residents before the flood would persist for decades.

In the flood's immediate aftermath, to ease the grief of those mourning the dead and disappeared, a requiem mass was celebrated at San Fernando Cathedral on Monday, September 20. The Right Reverend Arthur J. Drossaerts, bishop of San Antonio, officiated. In the cathedral's long central aisle, a series of tall candles surrounded a black pall-draped bier, a symbolic reference to the dead. "Requiem aeternam dona eis, Domine" were the first words of the service: Grant them eternal rest, O Lord. As shafts of light streamed through stained-glass windows, the high vaulted ceiling echoed with the "sweetly plaintive" notes of Gregorian chants, sung by the girls' choir of Saint Joseph's Orphanage. Chopin's funeral march followed, performed by the military band from Camp Travis (their presence was appropriate: hundreds of soldiers from the base had conducted invaluable search-and-rescue operations during the disaster). With Chopin's somber notes still hanging in the air, families and friends walked silently to the chancel rail to take Communion. In his formal remarks, the bishop thanked the city's many individuals and organizations for the "quick relief accorded the Mexican homeless," before offering consolation to "the sorrowing" and prayers for the "salvation and rest for the souls of the flood victims through eternity."[51]

The bereaved would not soon forget the Zepeda, Hernandez, and Morales families; the Garzas, Gittingers, and McCrarys; or the Cardenas children, among so many others. The Fraustos demonstrated their enduring grief in naming subsequent progeny after Ramón and Hortencia.[52] A collective anguish was indelibly etched into the community.

But for others, the flood's fatal force, its reverberating trauma, quickly faded. In a 1924 article discussing plans to clear away trees and brush growing in the narrow channels of the San Pedro and Alazán Creeks, the *San Antonio Light* gave this succinct rationale: "Several lives were lost when these two creeks went out of their banks during the 1921 flood."[53]

-{ TWO }-

RESCUE MISSION

On September 8, 1921, James L. Fieser, an alert general manager of the Red Cross's southwestern headquarters in Saint Louis, spotted the potential threat that a tropical depression posed to central and south Texas. He immediately communicated his concerns within the office and to national headquarters in Washington, DC, prepping leadership and staff of the potential demand on the agency's services. He also telegraphed chapters in the region, warning them of the possibility of inundation and encouraging them to make an inventory of necessary emergency supplies, especially food, clothing, tents, and medicine, and to start considering what financial support they might need to fund their relief operations.

It paid to be prepared, even if, on first appearance, his preparations seemed premature. W. Frank Persons, vice chairman of the American Red Cross, was initially skeptical. In response to Fieser's cautionary message, he fired off an early morning telegram on September 10 stating that the Washington newspapers had reported "there has been a flood in San Antonio" but noted they did not mention the kind of catastrophe Fieser had feared. "I suppose you are on the job and am glad that you have been saved a hurricane disaster as you anticipated might occur."[1]

Although Persons did not know it, his understanding of the situation was already outdated when he wired his colleague. Events on the ground proved every bit as tragic as Fieser had anticipated, a reality he began to relay to his superiors in Washington, staff in Saint Louis, and relevant chapter heads as soon as the Associated Press reestablished its wire connection with the beleaguered San Antonio. So rapid-fire were Fieser's communications via telegram and telephone, a reflection of the fluidity of the situation, that Persons felt compelled for archival purposes to amend his original (and inaccurate) message to Fieser. He scrawled on the bottom of it: "This was dictated long before our phone conversation."[2]

The trove of internal Red Cross documents like this located in the National Archives—including telegrams, letters, reports, and post-flood assessments—reveals the kind of miscommunications that can happen amid a rapidly evolving crisis. But these primary sources also offer important insights into the Red Cross's organizational abilities and emergency response efforts. As well, this documentary evidence captures the dynamic interplay between the national body, its chapters, and the communities it served in times of stress. When read in conjunction with local newspaper accounts of relief efforts in San Antonio, it becomes clear that the Red Cross's provision of food, clothing, shelter, and comfort to the afflicted was not as straightforward as it might appear. The voluntary nature of its local chapter, for example, had its benefits and limitations, a situation that led to some friction between the leadership of the San Antonio branch and the more professional approach to crisis management that the national Red Cross had developed during the first two decades of the twentieth century. Also complicating the task of helping those in dire need and with the recovery process were the social prejudices, economic disparities, and booster politics that did so much to define public life in San Antonio. For the city's Anglo elite, who were the central contributors to and volunteers for the Red Cross, it was essential to aid the city's poor in the brutal aftermath of the flood, but they had little interest in changing the conditions of the impoverished West Side. Relief had its limits.

There was no doubting the speed with which the local Red Cross chapter responded to the devastation. This was in part a reflection of its attending to the advance warning that Fieser had broadcast the day before the flood. That quick reaction in turn was a partial consequence of the conscientious leadership of local chapter president Albert Steves Jr. The scion of a wealthy and well-connected family, with the good fortune to oversee a strong chapter, Steves had the confidence of the national Red

Cross. "You will be interested in knowing," Fieser wired Persons, "that Mr. Steves is a comparatively young man who has been keenly alive to Red Cross interests from the beginning. He is one of the most alert chairmen that we have."[3]

Many of those who worked with Steves also had deep ties to the national body and considerable experience in relief work. "The Home Service Secretary, Miss Mabel Ferguson, is one of our most able workers," Fieser noted, as was field director Albert Shaw, who was "among the most experienced disaster workers in the Southwest." Another boon was that Shaw already was on "exceptionally good terms" with Major General Dickman at Fort Sam Houston, so that if "martial law or enlarged responsibility of the Army should eventuate, Mr. Shaw will be in a strategic position with that type of cooperation." These three individuals, Fieser felt, were perfectly situated to uphold the Red Cross's efforts, interests, and reputation.[4]

There was, after all, a brand to protect, a result of the American Red Cross's dramatic growth during World War I. Two years before the United States formally entered the conflict, in April 1917, the Red Cross had twenty-two thousand members. By war's end its membership had soared to twenty million. Those members' dues enabled the organization to raise more than $400 million to support its efforts at home and abroad, money that paid for a professional staff of more than twelve thousand and upward of twenty-five thousand trained nurses. These transformations came coupled with a radical shift in the national Red Cross's organizational structure. President Woodrow Wilson had appointed a Red Cross War Council to professionalize its leadership: it ousted the prewar leadership and replaced them with "wartime managers" who built the American Red Cross "into a corporation-like enterprise with fourteen regional divisions." The war gave the Red Cross a new "institutional identity," Marian Moser Jones observes, one that was "firmly embedded in the American psyche as the iconic symbol of comfort in crisis."[5]

Defending the organization's symbolic status fell to the now numerous "regional and departmental managers," many of whom, Jones asserts, "saw themselves as primarily responsible to the organization, rather than to the people it served." James Fieser gave voice to this perspective when he paused amid his tireless efforts on San Antonio's behalf to link relief work there with Red Cross interventions after spring 1921 floods that had roared through Pueblo and La Junta, Colorado: "There Italian and Slav and Mexican and native American gained courage alike as they saw the Red Cross in Action." The Red Cross had also offered succor in late May and early June 1921 when white vigilantes assaulted the African American community in Tulsa, Oklahoma: "In the midst of bloodshed and fire the emblem of the Red Cross had right of way." Fieser made certain that this emblematic presence was replicated in San Antonio, as he noted in one of his first telegrams about the city's flood-relief efforts: "All canteens and relief depots known to public as American Red Cross relief stations. All donations which approximate fifteen thousand dollars to be disbursed by Red Cross. This point given wide publicity. Under Red Cross leadership over three hundred volunteers working." Reinforcing this public under-standing of the Red Cross's centrality were the many photographs Fieser requested and then had published in the local and national press; clearly visible in each was the organization's ubiquitous logo.[6]

Fieser's confidence in the local chapter's ability to further the ends of the national Red Cross was not misplaced. Early on Saturday morning, before the full extent of death and damage was known, Shaw and Steves established their headquarters in the Chandler Building, located at the corner of Crockett and Losoya Street on relatively elevated ground near the Alamo; within a day, the *San Antonio Express* observed, it had become "a hive of industry." After conducting "a hurried survey of direct condi-tions," however, the havoc on the city's West Side convinced Steves and Mabel Ferguson that the Chandler Building was too far from the major devastation to serve as a "central relief station." Steves immediately req-uisitioned the city's Market House, in Market Square west of San Pedro Creek, placing the new headquarters in close proximity to some of the most battered neighborhoods and desperate survivors. "By organizing their forces early in the day," the *San Antonio Express* cheered, "the workers were ready to begin serving the homeless, the clothesless, and the starving refugees from the heavily flooded districts of the city."[7]

The Red Cross headquarters was located in Market Square.

The need was intense: "In the back of market place in the upper story of the market building there were two steady streams coming and going," the *Express* reported. "The entering stream came with practically nothing, not even hope. They left loaded down with clothing and carrying food for those unable to come to the square." Other victims stayed behind, bunking down in the building's second-floor auditorium, in which a kitchen and sleeping quarters had been hastily constructed. "The refugees, crowding in by the dozens, were given sandwiches and fresh water. They appeared utterly worn out, dazed, and helpless. They went wearily to the cots assigned them, and at once lay down."[8]

This snapshot dovetails with the final tallies of those served, data that is a striking testament to the flood's disruption of daily life and to the expanding size of relief operations. Cruz Azul, a West Side mutual aid society led by women, set up kitchens and distributed clothing in advance of the Red Cross's similar outreach, and in time would take over some of the Red Cross's commitments; it was aided by other *mutualistas*.[9] The

Salvation Army and the American Legion also pitched in. But there is no gainsaying the size and extent of the Red Cross's immediate responses. "During the emergency period of five days following the flood," upward of two hundred volunteers handed two changes of clothing to "approximately ten thousand men, women, and children." That amounted to more than eighty-five thousand garments, at least seventy thousand of which "were donated by the people of San Antonio and near-by towns." As for clothing for infants, "groups of women gathered in their homes, supplied the material and made these tiny garments ready for distribution."[10]

Those without clothes also needed to eat, and an earlier version of Meals on Wheels supplied the destitute with two meals a day. The US Army, through a cooperative arrangement with the Red Cross, provided and staffed seven "rolling kitchens" at key intersections on the West Side. In the afternoon they handed out sandwiches, and in the evening they provided roast beef, bread, and coffee, feeding hundreds of people a day, a remarkable operation.

To give this data a human face, local reporters made a point of chronicling the daily distribution of food, much as they would do a decade later amid the similar privations resulting from the Great Depression. In each case, journalists appealed to their well-heeled and literate audience, identifying the crying need for provisions, and in doing so also underscored the ethnic, racial, and class dimensions of the relevant disaster and affirmed the social distance between those in need and those who gave to those in need.[11]

One account in the *San Antonio Express* stands in for many others. While this report on the Market Square food distribution extolls the efficiency of the military's operation, it is focused not on the soldier-cooks but on the hungry, and employs less-than-subtle discriminatory language to describe their plight and peoplehood—what Natalia Molina describes as a racialized script.[12] A quick tally of those waiting in line amounted to several hundred children "all dark-eyed, all patient with the stoicism of their race," and their docility was reflected in their silence. "Generally hungry children cry. These did not. When anyone, even a child, is hurried from a home in the early dawn, their homes and all they valued swept away in

RED CROSS REPORT ON FOOD DISTRIBUTION STATIONS

Saturday, Sept. 12, to Wednesday, Sept. 21, inclusive

Tent City—Frio Street

Tom Hogg and C. D. Cannon
2 ARMY ROLLING KITCHENS

Opened Tuesday, September 13, to date	8 Days
Estimate of hot meals served	4,400

El Paso Street—West of Alazán

Tom Hogg
1 ARMY ROLLING KITCHEN

Opened Saturday, September 10; closed Tuesday, September 13

Estimate of hot meals served	2,400
Estimate of hot sandwiches served	1,100

El Paso Street—East of Alazán

Robert Harris
1 ARMY ROLLING KITCHEN

Opened Monday, September 12; closed Wednesday, September 14

Estimate of hot meals served	1,350
Estimate of hot sandwiches served	2,200

South Laredo Street—Rubiolo Store

Mr. Venne
1 ARMY ROLLING KITCHEN

Opened Saturday, September 10; closed Wednesday, September 14

Estimate of hot meals served	1,400
Estimate of hot sandwiches served	1,700

Market House

Opened Saturday, September 10; closed Tuesday, September 13
Estimate of hot meals served   6,000
Estimate of hot sandwiches served   11,000

Alazán Creek and I&GN Bridge

A. A. Foote

1 ARMY ROLLING KITCHEN

Sandwiches served Monday, Tuesday, and Wednesday
Kitchen opened Thursday, September 15; closed Saturday, September 17
Estimate of hot meals served   3,500
Estimate of sandwiches served   6,000

Josephine Street

D. K. Harris

1 ARMY ROLLING KITCHEN

Opened Tuesday, September 13; closed Thursday, September 15
Estimate of hot meals served   750
Estimate of sandwiches served   800

Cass Street Station—Alazán Creek

Miss Muller
Opened Sunday, September 11; closed Thursday, September 15
Hot coffee and sandwiches served   1,600

Grant Street Station—On Alazán Creek

Opened Sunday, September 11; closed Thursday, September 15
Served hot coffee and sandwiches to about   1,600

NOTE: Flood Relief Report of the Bexar County Chapter, American Red Cross to the Honorable Mayor and the Citizens of San Antonio, Texas (San Antonio, TX: Alamo Printing, 1921), NARA.

The army provided a number of rolling kitchens on the West Side.

a torrent of utter darkness, a person, even a child, does not cry or whimper. When the refugees from Belgium poured into North France or from the great Russian retreat in every section there were few tears; they had [passed] the time of crying. So with the Mexican children who ranged from [two] years of age. Patiently they waited until it was time to eat."[13]

As the air became redolent with the aroma of cooking, the "hungry ones formed in line," and those who then filed past the reporter did so in a certain order. "First the children were fed, then women, next the aged men, and finally able bodied. Perhaps 'unworthy' were fed. None worried. In a time of great disaster, one cannot investigate credentials. Those who came uncomfortable departed comfortable, at least for a time."[14] Left unsaid in this account, yet nonetheless visible in its language, was a set of presumptions about the stark differences between those who were being observed and written about and those who did the observing or reading. The flood victims' patience was perceived as a virtue by those who had not been chased from their homes, who had not lost everything in the flood's churning surge, for those whose status in and worth to the community were not subject to question.[15]

A similar constellation of assumptions would crop up in efforts to supply housing to those who had none. The first post-flood nights were the most difficult. Finding a dry and safe place for thousands of displaced people to sleep was resolved only by the voluntary actions and open-door generosity of Catholic churches and other houses of worship, as well as the Salvation Army, several hotels and fraternal organizations, and city facilities. "We housed every case that came to our attention," Steves noted, though he was unable to estimate how many people had been sheltered in this ad hoc manner. "The exact number of which is impossible to determine," he advised Mayor Black, "as you can well appreciate that it was a 'helter-skelter' performance, as the morale of the people who were homeless was rather at a low ebb."[16]

The survivors' psychological vulnerability, when combined with the precarity of their housing, led some to take matters into their own hands. By Monday, September 12, the displaced had begun to drift back to the low-lying neighborhoods they had fled two nights earlier. "Many of those whose homes were washed away have exhibited a desire to erect temporary shelters on the sites of their wrecked homes," the *San Antonio Express* reported—a nesting behavior that was eased ironically enough by the "great masses of wreckage" the flood had pile-driven into bridge abutments and other immovable objects or left scattered along the creek banks. Out of the tangled debris, the returnees salvaged enough lumber and boards to construct crude shelters, "'lean-to[s]'... on the ground where the victims' homes formerly stood." In this work they gained an important level of communal support from their more fortunate West Side neighbors. Those "whose homes are standing are aiding their fellow countrymen, and it is believed that many will attempt to remain in the flood-swept territory."[17] Surrounded again by family and friends, close to their parish churches, these families were attempting to reclaim their lives and their place.

It was the condition of these places that led the city to look askance at this self-help recovery operation. Public officials—in particular, the health

The Red Cross hired local labor to erect a tent city on the West Side.

department—argued that the returnees were creating a potential threat to public order and wellbeing. Local physicians were fearful of the unsanitary nature of these ramshackle abodes, which were being erected on muddy land still smeared with effluent, a landscape that lacked clean water and sewage infrastructure. Hoping to forestall an epidemic of such water-borne illnesses as typhoid, the city, with the support of the Red Cross and the US Army, decided to construct a tent city to better serve the homeless. With the initial goal of housing at least four hundred people, a projection that quickly swelled to two thousand living in five hundred tents, the Red Cross selected a site at the cross streets of San Luis, Medina, and Frio with preexisting sewer accommodations (latrines) and water access.

Although the tent city's stated purpose was to be "used by every section of the city where people have been made homeless by the flood," its West Side location signaled who the Red Cross anticipated its residents would be and why and how they would be moved to the new camp. "The work of transferring the dazed refugees, who in many cases, particularly among the Mexicans in the corral district, are still clinging to the relics of their homes, will be turned over to the County Health Officer, Dr.

Tents were arranged in a military-camp grid.

D. Barry, and his assistants." To facilitate this mandatory evacuation, a contingent of Red Cross social workers met with the afflicted to persuade them of the necessity of moving to the camp. Persuasive, too, was the decision to close many of the food distribution stations that had been located across the West Side and to centralize this service within the tent city itself. If people refused to move or disputes arose about their continued presence in the affected neighborhoods, the "authority of the sanitary inspector" would be invoked and when necessary, "the power of the law" would be "used to evict them from their flooded homes."[18]

The tent city never served the two thousand people its organizers originally anticipated, and as a result initial plans to develop three camps were made moot. These three camps would have followed the color line, providing separate accommodations for whites, African Americans, and Mexican Americans. As it was, the single tent city's mission was to act as a halfway house, a place of respite that would give its residents time to get back on their feet. Critical contributors to their restoration were the same Red Cross social workers who had escorted the first wave of residents to the tent city. For the next several weeks they remained on site "to find places of work and shelter for these people," a process of rehabilitation that Albert Steves hoped would boost the flood victims' morale and be of a "lasting nature."[19]

The other goal behind the construction of this temporary camp, and why it never housed as many as had been predicted, was the perceived need to reduce dependency on the city's relief efforts. "We tried to keep [the tent city] from materially increasing," Steves noted in his final report, "by taking care of only such cases as the Board of Health … wanted us to take care of."[20] To achieve that end required a different kind of intervention. At the same meeting where the Board of Health authorized the construction of the tent city and simultaneously required the removal of those living in the flood-wracked neighborhoods, the sanitation department announced it was spreading lime over the "filth and debris" the floodwaters had left behind. Its actions would be reinforced by another mandate: "Owners and proprietors of the corral or jacal district will be required by the city to thoroughly clean and disinfect" their properties and make all necessary repairs. None of the tenants would be able to leave the tent city, return home, or, not incidentally, pay rent until their domiciles "meet the requirements of sanitary inspector."[21] As each structure secured the requisite permit, another family decamped. By early October, the tent city had completed its mission.

In the meantime Albert Steves was drafting his final post-flood report, and he found much to praise. Of special note were "the splendid men and women who untiringly and unselfishly gave their time to this splendid work." It was, he argued, "work that anyone can be proud of if he had any part in the rehabilitation or emergency." And those who leaped into action taught him an important lesson: "I never knew that there were so many different kinds of people anywhere, as this flood relief showed me. It really was an education and a privilege to have conducted this work." It was this pronounced volunteerism that set the community apart: "Owing to the fact that this was the only disaster relief job that was ever handled direct[ly] by a Chapter, outside of the Division Office, or by the National Headquarters," Steves concluded, San Antonio's flood-relief commitment was "in a class by itself."[22]

As Steves's congratulatory analysis suggests, the flow of aid and effort seemed seamless. The local and national Red Cross, US Army, city and county, and numerous civic organizations appeared to be acting in unison, offering a textbook example of cooperative emergency management. But a series of backstories challenge some important elements of Steves's rosy account and highlight complexities in San Antonio that may have diminished the community's ability to respond fully and effectively to the crisis.

One of these backstories begins with a telegram. Anxious that the nation not get the wrong idea about San Antonio's capacity to recover and rebuild on its own in the flood's aftermath, on Sunday morning, September 11, Mayor Black wired leading newspapers around the country, including the *New York Times*: "Condition in San Antonio is exaggerated. The loss of life is less than fifty. The property damage is about $3,000,000. The city is able to care for itself and does not need outside help. In two days all will be back to normal."[23]

The mayor had a point. The wires, based on early reports, had sensationalized the death and destruction, with headlines like the one in the *Grand Forks Herald* ("Flood Takes Fearful Death Toll; 100 Believed Dead in San Antonio"), the *Rock Island Argus* ("San Antonio Is Swept by Fatal Flood; Lives Lost May Total 500; Wrecked Houses Strew Banks of Rivers," and the usually staid *New York Times* ("40 Known Dead, Fear 250 Perished in Flood That Sweeps San Antonio; Property Loss Is Put at $3,000,000"). Black's telegram, at once terse and reassuring, was designed to calm an overeager press and prop up the city's reputation. Battered, it would rise again by dint of its own efforts.[24]

Yet Mayor Black's assertions were inaccurate. Life would not get back to normal in two days or two months. More striking was his declaration that San Antonio could and would clean up its own mess and that the community would recover without outside assistance, full stop. Even as he dictated those carefully chosen words, Black knew they were not true. As soon as floodwaters began to churn through residential neighborhoods and the downtown business district, the US Army had dispatched men and material from Fort Sam Houston, Kelly Field, and other local bases in a rapid and full-scale intervention that was critical to rescue operations in the first two days. The mayor knew too that aid had flowed into the city from across the state. Houston had dispatched two thousand cots, purchased from the US Army, that were later assembled in the tent city. The metropolis also sent five thousand loaves of bread, a vital staple that other Texas towns and cities also contributed on a daily basis. Then there were the manifold contributions of the American Red Cross. Its professional staff had arrived from as far as Missouri, Colorado, and the nation's capital. As important was the Red Cross's $20,000 contribution for direct relief and another $5,000 for administrative expenses; in the face of crisis, the organization served as a much needed and stabilizing outside force.

The local press was perfectly aware of the array and significance of these and other external donors and their support, and routinely praised these examples of philanthropic generosity. Yet in concert with the mayor and chamber of commerce, the *San Antonio Express* and its rival, the *San Antonio Light*, helped promote the notion, in the mayor's words, that "San Antonio is able to care for itself." Learning that national Red Cross official James Fieser and a number of Red Cross social workers were on their way to the city "to take charge of the work," the *Express* offered a backhanded *bienvenidos*: "Any assistance that may be received, even from the moon or Mars, will be welcomed, but it is not needed. San Antonio can care for its own."[25]

Caring for itself—the slogan resonated with a local political goal to appear self-reliant and independent. Keeping up the appearance was crucial to the coalition of Anglo downtown business leaders who had run Mayor Black's campaign in spring 1921 under the banner that he was the "anti-administration" candidate, by which they meant he would oppose the political machine that drew its support from working-class neighborhoods like those disrupted by the flood. In a hotly contested race, Black squeaked by at the polls in May and was thus beholden to those who had funded and managed his candidacy. No surprise, the mayor tapped these same individuals—among them leading bankers and other wealthy business and civic leaders—to organize the finance-relief committee.

They were very good at this task, securing $25,000 within two days and later doubling that amount. This financial outlay seemed to bolster San Antonio's claim that it needed no external aid, a perspective local media reinforced as a matter of civic duty and pride: "I am the Spirit of San Antonio," one headline cheered. "I will fight all the harder to make San Antonio the flower of the southwest, to bloom courageously...in the wreckage."[26]

Tough talk notwithstanding, behind the scenes the finance committee was operating with a different set of calculations. Even before Mayor Black asserted that the city would seek no outside help, committee members anticipated the impact this might have on the Red Cross's generosity. To preempt withholding of funds, while stoutly maintaining their public denial that the city needed aid, the committee used local newspaper contacts to release a story announcing that the national Red Cross had promised to donate $20,000 to flood relief. When an embarrassed James Fieser queried his superiors in Washington, DC, about why he had not been told of the announcement, Vice Chairman Persons telegraphed that the donation was news to him, too. "Absolutely no statement issued to press from headquarters concerning grant of funds from Red Cross.... Probable that those [on the local relief fund committee]...have occasioned such press statements." It is also likely that these same individuals blocked publication of the Red Cross's expressions of support that Persons had cabled to local chair Albert Steves on Sunday morning: "On behalf [of] national Red Cross Organization permit me to express our deepest sympathy [to the] people [of] San Antonio in their hour of trial and to offer them...any friendly assistance for which you may call on us." Persons noted: "Even this was withheld from the press."[27]

The Red Cross, however manipulated it might have been, formally acknowledged that it would donate $20,000 to aid in the relief of flood victims. But it drew a hard line against sending additional funds to a community whose leadership asserted that it needed no outside support. In a lengthy telegram on September 13, James Fieser informed the national leadership that city officials were working against their own best interests. Local chairman Steves reported that there had been a "serious loss to

Business district already economically stricken," a blow he acknowledged would impact the success of any "local appeal" for rescue and rehabilitation funds. As the mayor was announcing that the city could go it alone, Steves was privately reporting that additional funds from the national Red Cross would be essential to help with the recovery. "I have no additional word from Steves to date," Fieser wrote, "but am wiring him that he must correct press impressions issued in the name of the chamber of commerce and city officials that outside help is not needed if San Antonio expects much additional outside help."[28]

This point was reiterated over the next couple of days: each time Steves appealed for more money, the national organization reminded him that city officials had declared they did not need it. "The Red Cross will not ask for more funds for San Antonio from public until such appeal is sanctioned by city officials and local chapter and publicly approved by them." Another internal Red Cross document noted bluntly: "We cannot allow our contingent fund to be raided by the city when they are publishing broadcast a statement that they do not require outside help."[29] Even if San Antonio were to change its tune, Fieser warned Steves, it had probably lost its best opportunity to capitalize on a nationwide appeal. News of its devastation, which could have encouraged an often generous American public to donate to the city's relief fund, was beginning to wane: "San Antonio flood news now off front page newspapers," he wrote. That was even true in the city itself: one week after the flood, local media were pushing follow-up stories about the crisis to their interior pages. The "psychological moment," noted Fieser, had passed.[30]

It may have passed, but the need remained intense. No one knew this better than Albert Steves, which is why well into November he privately pleaded for more money from the Red Cross despite the city's continued refusal to publicly request support. A sympathetic Red Cross nonetheless insisted that it would not send more funding until San Antonio went public with its need. Fieser dialed up the pressure. "The Executive Committee of the American Red Cross," he wrote Steves, "is composed for the most part of business men, all of whom have been impressed, as have Mr. Hoover and President Harding, by the oft-repeated statement that

San Antonio did not need outside help and would take care of itself. They also recall that the Washington office was ready and in a position to undertake to secure very substantial additional sums at the time the flood took place if permission were given them to do so."[31] The troubled relief effort was the city's own doing.

In the background, the chamber of commerce and the local finance-relief committee, with the backing of Mayor Black, adopted a different strategy for reaching out to the nation; this tactic may explain why they resisted acknowledging the flood's continuing impact on the city. The chamber printed fifty thousand copies of "The Truth about the Flood," a circular detailing how quickly the city had repaired streets and bridges and how fast downtown businesses, banks and hotels included, had re-opened (within twenty-four hours). Although the total financial loss of $5.6 million was significant, the leaflet noted, the city would never again suffer such a costly inundation, for already borings had been made in the Olmos Basin for the "immediate construction of a masonry dam for flood prevention." Historic missions, like local golf courses and parks, were uninjured, a nod to what the chamber hoped would be a quick recovery of tourism to pre-flood levels. There was no hiding the fact that people had died, although the circular's tally of fifty-one deaths was an undercount. But that, too, served the document's purpose, which was to rebut what it called "grossly exaggerated reports" about the damages the flood had inflicted on San Antonio and that were spread across the nation "by certain news agencies." To circulate the rebuttal, the chamber called on local commercial concerns to mail copies to their suppliers and customers, utilizing these informal networks to change the narrative.[32]

Seeking greater exposure, the chamber also encouraged its membership to attend all relevant regional and national conferences. One such gathering was the 1921 meeting of the Texas Association of Real Estate Boards, attended by a large delegation of San Antonio realtors. By their presence and voice they promoted the idea "that the Alamo City has 'come back' after the flood,"[33] a boosting message that the chamber and its allies also conveyed in high-level meetings and tours of the city with the state's lieutenant governor and mayors of major cities. Somehow Mayor Black,

who was all-in on this campaign, muffed one element.[34] As he squired his
political peers from Fort Worth, Houston, Dallas, and Galveston around
San Antonio to demonstrate how quickly the city had been rehabilitated,
he undercut that claim by announcing major budgetary cuts to the health
and parks departments. City commissioner Ray Lambert was baffled by
this decision at such a critical moment "when it appears most important
that the city maintain its reputation as a health resort."[35] As if to paper
over the mayor's mistake, the chamber of commerce took out a major
advertisement in the *Saturday Evening Post,* one of the country's most important magazines, not to seek help for the city's displaced residents but
to boost the local tourist economy. The advertisement's reported funding
source shocked Red Cross officials, Catherine Fennelly notes, as did its
purpose: a "large part of the money collected within the city" for post-
flood relief, reconstruction, and rehabilitation was diverted instead to
"advocating San Antonio as a vacation resort."[36]

Such schemes were not the only issue to surface when Red Cross officials debriefed about the organization's South Texas relief efforts. "I
think you and I are entirely agreed on the San Antonio question," Edward Stuart, director of the American Red Cross Disaster Relief Service,
wrote Fieser in late November, "namely that the whole set up there was
not entirely satisfactory from our point of view."[37] One critical dilemma,
structural in character, emerged in the wake of the controversy over donations. That the city's relief-finance committee had gone over the head
of local Red Cross chairman Steves to lobby the national leadership for
funding highlighted the lack of strict parameters for communications and
decision-making within the organization's professional staff, and between
them and local volunteers. James Fieser had spotted this problem shortly
after receiving news about the flood, writing his superiors in Washington that San Antonio already offered considerable "evidence that disaster
work cannot be administered from two points at the same time."[38] A chastened Vice Chairman Persons agreed, responding that "the whole moral
of this letter is that we both agree that responsibility must be centralized
in a tense and emergent situation" and that henceforth "you may count
upon me always to have that in mind."[39] He made good on his promise,

agreeing that in the future "we should approach a chapter with a clear-cut proposal for our taking over the work once outside help is needed." If aid was deemed unnecessary, then "we should definitely leave it to them, acting in an advisory capacity if they should desire it, but without our taking any of the executive responsibility."[40]

In San Antonio, this administrative dilemma was compounded by the decentralized nature of Red Cross activism, which depended heavily on local chapter leadership, volunteers, and community goodwill to institute its relief missions. There were limits to this organizational model. So noted Red Cross official Arthur Shaw, who arrived in San Antonio from his post in El Paso on September 12 and immediately took charge of overall operations. Cheered that "all chapter workers, camp workers, and volunteers working day and night [are] doing well" despite the dire situation, he spotted how much overlap of labor and waste of resources were occurring and the resulting impact on the delivery of critical services to flood victims. More worrisome was that the conditions on the ground appeared to be deteriorating, and despite what he had written about the strength of the volunteers' morale, he noted that those men and women who were pouring all their energy into stopgap measures were beginning to burn out and drop out—a diminishing of the labor pool that led Steves just two days after the flood to plead for replacements to staff relief stations and to drive food over to the West Side. It was in this context that Shaw telegraphed Fieser seeking an immediate influx of professional staff: "Surely need help in way of trained disaster personnel and principally man with experience in properly organizing general disaster work."[41]

Conceding this point was Albert Steves. In his final report on the flood-relief efforts, he acknowledged that strictly relying on volunteers may have hampered the local chapter's outreach to those in dire straits. Although volunteerism "had its compensations...at the same time it is, of course, left for the public at large to state how well the work was done, and whether it could have been done better by professionals altogether."[42] For the American Red Cross, the answer was clear: without its professionals and their expertise in organization, logistics, operations management, and social work, the crisis would have reached catastrophic proportions.

The pressing need for its expertise, Red Cross officials believed, was also evident in the city's less-than-robust plans for the rehabilitation of the West Side neighborhoods that the flood had flattened. The question turned on the quality of homes to be constructed, as well as who had the power to make those decisions and on what basis. The city's objective was to rebuild the so-called "corral district" as cheaply as possible, an approach the local finance committee—some of whose members were landlords in the devastated region—fully favored. With underwriting support from the relief fund, the return on investment for these landowners would be considerable. Stated differently: Why spend a lot on shelter for those who had so little? That query, which guided the finance committee's actions, was itself framed by the racism of the city's Anglo leadership and its longstanding disregard for the large and growing Mexican and Mexican American population crowded into substandard housing.

No wonder, then, that the local chapter and its requisite committees chose to ignore the Red Cross's public challenge that J. B. Gwin, who had managed the national organization's intervention for the first month, published in the October issue of the *Survey*, a progressive magazine. The city "has not shown the least interest in preventing a recurrence of the bad social conditions among the Mexican population, which were crying for remedy years ago." This was too bad, Gwin wrote, because the "destruction of these shacks and even whole corrals has given an opportunity for bettering the lives and sanitary conditions of the Mexican population." The first step, Gwin had counseled Steves before he departed the city, was to undertake "a complete survey covering not only present conditions, but the actual loss in property and homes, which were either completely or partially demolished." Had that data been gathered, he wrote, had it been conducted with the aid of "Spanish speaking workers or with the help of interpreters," San Antonio would have had access to much more complete information to guide its subsequent decisions about where and how to rebuild. That the city refused to conduct any such interviews is telling: officials had no interest in learning what those on the West Side felt would be in their best interest, a disdain that quite literally was hammered into the first twenty-five houses the local Red Cross chapter's salvage committee had erected by late October.[43] At $40 a house, the chapter had spent a total of $1,000 on these structures, and their cost reflected their quality.[44] "I made an inspection of the results," wrote Henry Baker, who had replaced Gwin as the American Red Cross's representative in San Antonio, "and regret to report that the houses are not such as we could desire." This criticism led him to recommend to Steves "that from now on any building which is done shall be done by the Red Cross alone and not in cooperation of the Salvage Committee." This recommendation, and by implication the criticism, Steves "very heartily accepted."[45]

He did not fully embrace Baker's perspective, though. In a letter to Fieser, Steves conceded that "there is a great deal of argument about the job being done cheaply. However, I feel confident that we would have heard a great deal more complaint if the people as a general thing were dissatisfied."[46] How Steves was able to ascertain the people's level of satisfaction without carrying out the survey Gwin had proposed is open to question. His response to Baker's critical analysis of the rebuilt homes' poor quality demonstrated a disregard for the very people the Red Cross was there to serve. "Mr. Baker makes the big point that the housing had been so badly taken care of. This is no doubt true, Mr. Fieser; but we aided people wherever possible."

A defensive Steves then derogated those living on the city's West Side: "A great many of the houses that were utterly ruined, were valueless anyway, and the Mexican families which had inhabited them might just as well be living somewhere else as in San Antonio and no doubt they have moved before now." Besides, he wrote, "it is impossible to rehabilitate people as before the flood, because it would not be fair on the donors of the money itself." Prioritizing the expectations of the philanthropic few over the devastated many, Steves rebutted the Red Cross's conviction that much more money should be spent on new housing, as Baker had urged. "I believe that Mr. Baker's housing problem, although he seems to solve it in a very splendid way, does not altogether mean a necessity as Mr. Baker may see it." Steves was certain that locals saw things differently, that the "citizenship here is sufficiently satisfied with the work that the Red Cross did."[47]

Citizenship. Here, then, was use of one of an interlocking set of rhetorical devices that the San Antonio elite deployed to determine who was a full-fledged member of the community and who was not. Who had rights

RED CROSS IDENTIFIED THOSE IT SERVED
BY RACE, RELIGION, AND NATIONALITY

	Number of families helped	Number of people in these families
Americans—White	605	1,479
Americans—Colored	79	261
Mexicans	1,966	6,736
Italians	7	38
Polish	I	10
German	15	43
Belgians	I	5
French	I	6
Bohemian	I	9

The terms and data in this table come from the Flood Relief Report of the Bexar County Red Cross Chapter (1921).

in the city, and who did not? In short, who belonged? "The citizenship" implied San Antonio's white residents. The term "Mexican," meanwhile, was an all-encompassing label white San Antonians used to lump together those living on the West Side, whether they were residents, longtime US citizens, or recent immigrants. They were the "other," segregated from and set in contrast to anyone who bore the label "American." Statistics on who was receiving aid during the flood crisis broke recipients labeled as "American" down into explicit subcategories along racial and ethnic lines. But for the West Siders, there was only the one designation: "Mexican."

This othering of the flood victims was reinforced in the local media's description of the survivors as "refugees" and likening them to those fleeing from Belgium into France during World War I—they were Mexicans, after all, not Americans.[48] The othering was also reinforced in the local Red Cross chapter's articulation of who was worthy of receiving support, who could tap the community's largesse. The answer, it turns out, was that individuals proved their worthiness by *not* taking aid. Steves developed this counterintuitive argument while explaining the local organization's rationale for offering only small sums to help the impoverished get back on their feet. "We have tried to be as fair in our awards as we could possibly be," he assured Fieser, but "I believe that the way we handled it here is far better than a lavish big splurge such as I understand was indulged in other disasters." The proof, he declared, was manifest in the list the chapter sent to "the Commander of 8th Corps Area of Fort Sam Houston of the people in Government employ who had suffered by the flood." Every case was "carefully investigated, and I would say that in 90% of the number, help was declined. The same obtains of people along Oakland St. and in the better section of the town where, of course, there were some people who accepted help, but the great majority of people did not."[49] By refusing public support, these residents earned public acclaim.

What Steves did not acknowledge in this calculation was that, unlike West Side residents who lost everything, including family members, homes, jobs, and income, the civilians working for the US Army remained fully employed. As for those wealthier, North Side whites, they, too, had an array of resources—financial, social, and psychological—that enabled them to cushion themselves from whatever losses they may have endured. To alter these unequal conditions and the spatial injustices that reinforced them would have required a wider, more inclusive conception of San Antonio than Steves and his peers were willing to embrace. Indeed, their narrow understanding of who and what constituted the city was further solidified in their universal demand for and unequivocal support of the construction of the Olmos Dam, a flood-retention structure that secured the downtown's economic fortunes but provided no relief to those living on the low-lying West Side.

Although residents of the district had to wait until the 1960s to secure even a slow rollout of flood control infrastructure, they gained more immediate political compensation. In 1923 their votes helped power county sheriff John B. Tobin into the mayor's office and by an equally decisive margin helped elect a very friendly slate of city commissioners. Black, who had decided not to seek a second term, may have stepped aside, candidate Tobin reportedly declared, because of his mishandling of the flood. Mayor Black, Tobin mocked, "was washed out of office."[50]

-{ THREE }-

MILITARY INTERVENTION

Splintered homes. Buckled bridges. Gutted streets. Collapsed buildings. Displaced and distraught adults and children. The stomach-churning stench of death. A flood-battered San Antonio, bewildered and reeling, felt war-torn, an analogy the *San Antonio Light* drew on in its first post-flood edition on Saturday, September 10, 1921: "About midnight, as it became apparent that the waters were still rising, a big truck rumbled into San Antonio and stopped short. A sharp command rang out and a detachment of khaki clad soldiers sprang into the streets, fixed bayonets and swung away in columns of twos for this flood area." Their mission was to protect the downtown business district from looters: "Within a few minutes they were doing sentry duty as though they had not been called from sleep suddenly to protect a city beset with death and disaster."

Moments after they set up their patrols, additional relief arrived: "a clatter of hooves sounded on the pavement, there was a splashing of water at the edge of the flooded area," and once again "a sharp word of command" rang out and the "mounted troops of Fort Sam Houston and Camp Travis went into action—not to take life but to save life." As these riders plunged into the raging waters, they pulled men, women, and children to safety. With their example, and bolstered by the presence of "soldiers on foot giving aid at every turn...soon civilians, soldiers, policemen, and firemen were intermingled in the great task of saving lives," a chaotic process that took place "in complete darkness, except for countless flashlights, smoking kerosene lanterns and motor car spotlights

Pavers litter East Houston Street in the flood's aftermath.

which gleamed and danced and flickered here and there...like dozens of big glowworms."[1]

When dawn broke, the rival *San Antonio Express* also sought to illuminate how embattled the city was, by relating its residents' harrowing experiences to the suffering that Belgian and French civilians had endured when the German army smashed across their nations' borders in 1914. "In the western part of San Antonio where flows the Alazán Creek through the thickly populated Mexican district," a reporter wrote, "there were enacted scenes of the flight of refugees before advancing armies when the World War was at its height. Only in San Antonio the flood victims had less warning than the booming of distant cannon."[2]

The comparisons were understandable. The armistice that had ended the murderous trench warfare of World War I had been signed less than two years earlier. Like many Americans, San Antonians were keenly aware of that brutal conflict, however far removed they were from those bloody European battlegrounds. They paid close attention to its vicissitudes as well because the city had a direct connection to that global conflict: as a staging ground, it had contributed significantly to the US war effort. Arguably the most militarized city in the nation, San Antonio was home to Fort Sam Houston, a sprawling facility that since its establishment in 1879 had been training soldiers sent to battle American Indigenous Nations out west, to exert US control over Caribbean islands, and in 1916 to give futile chase to Pancho Villa in hopes of stopping his cross-border attacks. These missions had led to Fort Sam's continued expansion, so that at the start of World War I it was one of the nation's largest posts.

No sooner had the United States declared war on Germany in April 1917 than the post exploded in size and operation. Its preexisting training facility, Camp Wilson, was renamed Camp Travis, but the far greater transformation occurred as construction workers swarmed over the camp, hammering together 1,400 temporary buildings in three months that would house, feed, and train more than a hundred thousand soldiers who served in Europe during the war.

The army's first airfield had been carved out of Fort Sam during the prewar years. Known as Dodd Field, this single aerodrome proved insufficient to meet the intensifying demand for pilots and ground crew in the first months of the war. In response, the Department of War leased seven hundred acres south of the city for what would be named Kelly Field and within twelve months increased its size by another eighteen hundred acres. As of Christmas 1917, more than 39,000 men were stationed at Kelly; even as the base shipped out 15,000 cadets the next month, another 47,774 recruits arrived. During "the hectic months of 1917 and 1918," Thomas A. Manning observes, "Kelly soldiers organized approximately 250,000 men into aero squadrons." The enlisted mechanics training department alone trained upward of "2,000 mechanics and chauffeurs a month. Most of the American-trained World War I aviators learned to fly at this field, with 1,459 pilots and 398 flying instructors graduating from Kelly schools during the course of the war."[3]

With Kelly and Dodd Fields overrun, the army developed a third airfield. On 1,300 acres, Gosport (later Brooks) Field trained pilots to fly balloons and airships (dirigibles). Other training facilities sprang up or swelled in size, including a rapid ramp-up of the army's medical services. Doctors, nurses, ambulance drivers, and stretcher-bearers all passed through San Antonio,[4] as did hundreds more assigned to the artillery. The ordnance they fired during training came from the local arsenal situated just south of downtown San Antonio, a facility whose storage structures and capacity swiftly expanded in the run-up to the war.[5] Regardless of where one lived or worked in San Antonio, it would have been impossible to miss the sights, sounds, and smells of preparation for the conflict. The Somme, Verdun, and Saint-Mihiel battles may have taken place Over There, but many US troops who fought had been trained Over Here.

By war's end, the city's spatial geography had become demarcated by wartime mobilization: military camps, cantonments, forts, and airfields were now the dominant cornerstones of the community's southern rim and northern extent just as they were the building blocks of the local economy. Some of San Antonio's streets also bore witness to its wartime commitments, emblazoned with the names of some of the conflict's decorated heroes—Funston, MacArthur, Mitchell, and Pershing. It is little wonder that residents trapped in the flood's fury felt such a profound sense of relief when they spotted hundreds of soldiers on foot and horseback coming to their rescue. Or that local media immediately drew parallels between the armed services' tide-turning contributions to the Great War and the Great Flood of 1921.[6]

Over the succeeding days that September, reporters were on the lookout for heartwarming signs of the army's engagement, which, it turns out, were around every corner. An estimated one thousand troops poured into the city, bringing food and water, setting up tents and portable kitchens, and providing other necessities for survivors stripped of life-sustaining essentials. Confronted with dozens of damaged or destroyed bridges, the soldiers, as in France, assembled pontoons. They launched a small flotilla of "German half-boats" whose crews plucked an estimated hundred people out of danger, and, tragically, forty bodies of those who had drowned. On drier land, military patrols intimidated any potential looters. There was a lot to praise.

Singled out for his behind-the-scenes management of men and matériel was Col. Leon Kromer, assistant chief of staff of the Eighth Corps Area. A graduate of West Point, Kromer had decades of experience in logistics and operations—he had been Gen. John J. Pershing's quartermaster during the 1916 Punitive Expedition into Mexico and had served on his staff during World War I with oversight of the flow of goods and services into and out of the war zone. Kromer brought that same close attention to detail to his command of the army's flood-relief response. "The experience in traffic control, gained in France during the world war, stood [him] in good stead when he took charge of the situation here last Saturday," the *San Antonio Light* affirmed. Kromer was praised "by all sides for his ready grasp of [the] situation, and the dispatch with which he, with the assistance of police officers, brought order out of a chaotic situation."[7]

Like Kromer, many of the officers and soldiers had served in the war, experience that aided the army's coordinated response in this peacetime emergency. But what really captivated local journalists, and presumably their readers, were individual displays of human courage outside the chain of command. Among the most lauded, at least in Spanish-language accounts, was Alfredo Gutiérrez. Stationed at Camp Normoyle, yet another installation born of World War I, he was touted as a gifted swimmer, a skill that became critical to his rescue work at the height of the flood. Gutiérrez was part of a detachment dispatched to South Flores Street, close to the roaring confluence of Alazán and San Pedro Creeks, scene of some of the worst destruction.

What struck the journalist who interviewed Gutiérrez about his exploits there was the young man's modesty: "He doesn't seem to be moved by the work that has been done, which has all the hallmarks of an epic." When asked how many people he had pulled from the torrent, Gutiérrez paused and "remained meditative for a moment and then answered with conviction. 'You know, it wasn't only me ... an American soldier I don't even know the name of helped me a lot. We took out some children first ... we pitied them with their childish voices. ... I would have liked to have had a thousand arms to hold them all ... they all cried ... but to catch some it was necessary to leave the others ... you see. ... God has given me only two arms. ... In short, I managed to get 18 out of the water.'"[8]

Even as Gutiérrez enumerated those whom he had saved—"Eight women, seven children, and three men"—he counted the many more he could not. The debris-laden, "muddy waters prevented me from doing the work of rescue....More could have been done...much more...but my strength, unfortunately, was spent more repelling the aggression of the sticks that unceasingly threatened to sink us, than in saving those who drowned." Haunted, he could still "see in front of my eyes women, whose arms were stretched towards me in frightful as well as imploring piety, passing...heartbreaking when with my strength and my time, it was not possible to save those who were drowning without employing more effort. And, pleasant, very pleasant, when I reached the shore and understood that one more being owed me his life."[9]

Perhaps as conflicted was the unnamed soldier who, a few miles away from Gutiérrez, had jumped into the surging San Antonio River to grab hold of those trapped in its tumultuous waters, an act of bravery that nearly cost him his life:

> By throwing a blanket from the window of her second-floor apartment, Mrs. E. Coffman, 222 East Pecan Street, rescued a soldier from the rising water of the San Antonio River at about 1 o'clock Saturday morning. The soldier was attempting to rescue several Mexican children in danger, when he was carried down the river by the force of the current. He caught a telephone pole a short distance from Mrs. Coffman's window and clung there until the blanket was thrown to him.
>
> When rescued, the soldier was in critical condition as the result of cramps incurred by his exposure. He is said to have saved the lives of two Mexicans before being carried away by the river.[10]

When Gutiérrez and his comrades-in-arms waded into the torrents, they embodied a selfless commitment to their duty and the community's safety.

The same held true for those manning the army's eight lifeboats. "These crews, in spite of the current and danger from floating debris and hidden snags," performed admirably, according to a military post-flood debriefing report. Many "people were rescued who might have perished, as other boats were not available." Even the one instance of clear danger ended fortuitously: "one boat hit a snag and was capsized," and although the boat itself was swept downstream, "all the occupants were able to make shore in safety."[11]

Soldiers ply the Saint Mary's Street floodwaters in a pontoon boat.

The military's swift and effective rescue operations earned the gratitude of civilian authorities. "The mayor and the Police-Fire Commissioner have stated that a police force, doing its utmost to send warning and give aid, was able to reach but a very small percentage who were endangered by the flood," the formal army report noted. The "immediate answer by the Military to the City's call for help resulted in the saving of an inestimable number of lives and the avoidance of greater property loss." Yet lurking beneath the surface of that praise were a number of metaphorical snags. The political crosscurrents that the army's search-and-rescue operations encountered proved as murky and treacherous as the overflowing creeks and rivers.

That the army was asked to intervene was an understandable consequence of the police and fire departments being overwhelmed. Yet perhaps a reason for what in retrospect seems like the local agencies' lack of preparation is illustrated by the embarrassing prediction the police department issued as rains fell that Friday. In a top-of-the-fold front-page article in the *San Antonio Light*, the city's afternoon daily, J. H. Jarboe of the US Weather Bureau had indicated that the downgraded but still worrisome hurricane already had dropped a substantial amount of rain. Indeed, it was close to the local record of 7.08 inches in twenty-four hours set in October 1913; that particular storm had unleashed a blockbuster flood in which "the river overflowed causing damage and loss of life." Jarboe then correctly forecast a very wet, record-breaking weekend.[12] -{ 81 }-

Despite this forecast, the police seemed unimpressed. Under what would end up being a provocative headline, "Police Expect No Flood," the department assured readers that "no dangerous rise in the San Antonio river" was anticipated. After all, it had required "a seven inch rainfall over two and a half days to cause the 10-foot rise in the river in 1913 and again in 1914," and the police did not expect history—even such recent history—to repeat itself. That said, a spokesman confirmed that the preceding day's rains already had damaged macadam-paved streets across the city and that many of the arterials and plazas hardened with heavy mesquite block pavers had "bulged," endangering pedestrian movement and vehicular traffic. Worse, on the western edge of the downtown core, so much water was flowing that "hundreds of pieces of the block pavement" were being swept down West Houston Street before tumbling into San Pedro Creek. Hours later this waterlogged debris—in combination with uprooted trees, crushed automobiles, and mangled homes and buildings—began to bulldoze bridges, pound downstream structures, and kill or maim those who found themselves ensnarled in the roiling water.[13]

The police department had little opportunity to reflect on its now seriously flawed calculations, because by that time its capacity to react was severely limited. Like a ship foundering at sea, the department's headquarters was taking on water by the middle of Friday evening, resulting in an oil- and creosote-smeared mess that destroyed years of records even as it forced staff to flee to the still-dry second floor. With electricity beginning to flicker and telephone communications becoming unreliable, a set of consequences ensued. The first was the disruption of the informal flood-warning system the police had devised after the 1913 and 1914 floods. In the wake of those damaging events, the police had asked ranchers and dairy farmers who lived in the Olmos valley to set up gauges to monitor rainfall amounts; if totals spiked above seven inches over a short period, they were to alert the city's emergency services. Yet in September 1921, malfunctioning telephones meant that no such warning came. When police and fire commissioner Phil Wright failed to make contact with Joseph A. Horn, a dairy farmer who lived on Blanco Road, he dispatched two motorcycle officers to investigate. They never reached their destination, returning to the city within thirty minutes to report to the commissioner that it had been "impossible to reach the Olmos valley because of extreme high water."[14]

Recognizing how quickly the situation was deteriorating, Wright sent messages out to all police officers and firefighters to report to key staging areas: at two sites along the San Antonio River—Brackenridge Park and downstream at the Josephine Street bridge—and at the Commerce Street bridge where it crossed Alazán Creek. Once there, they were to fan out and warn residents, where possible, to seek higher ground. How many officers received this command, and how many residents heard or heeded these warnings, is unclear, because by the time Wright learned of the impending disaster it had become impossible to centrally organize officers' response (or for the officers to act in concert with one another). "Practically the entire strength of the City Police force, 203 men, and fire department, 232 men, were on duty," the army's post-flood report indicated, and the majority were compelled to work "individually because high water made it impossible to reach headquarters."[15]

Wherever the flood found the city's first responders, there they stayed, rendering aid. Their individual initiative caught the attention of the *San Antonio Light*, which assumed there had been greater coordination than in fact had been possible. "Quickly marshaling their forces, the fire and police departments responded nobly and wholeheartedly to the call for rescuers. At every danger point, wherever people were appealing for aid, the blue coated figures of San Antonio's police and firemen could be seen, doing yeoman service without thought of personal safety."[16]

Their valor notwithstanding, Wright knew better than the media just how scattered and solitary his forces were, even before the flood reached its crest. To secure reinforcements, he drove through wave-washed streets to Fort Sam Houston, located on Government Hill overlooking the city to the northeast. There he hoped to enlist the army in the fight to protect the community. His was a smart move, given the military's command-and-control structure and its human and matériel resources. In short order, soldiers were called out of their barracks and loaded into trucks while mounted troops headed to the stables to saddle up. In the interim, a cadre of officers raced downtown to set up command posts at the post office, and at major bridges and transit routes. From these and other locations they would direct rescue operations and lock down the beleaguered city.[17]

The army's projection of power—badly needed and with immediate impact—was not uniformly embraced. One of the flashpoints concerned chief of police Al S. Mussey. Because martial law had not been declared, and would not be, military authorities determined that they must "work under and in close co-operation with some civilian official." When an unnamed army officer conveyed this request to Mussey, the chief "declined to accept the responsibility, requesting that the army officer in charge of troops take complete command of the situation." A second request received the same response, delaying the deployment of troops. Reporters later learned that the officer who was trying to convince Mussey of his duty became so frustrated that he urged Wright to temporarily replace the chief with "someone who was capable of handling the serious situation." Wright demurred: only the mayor could make such a change, but Mayor Black, having appointed Mussey to his position, was loath to replace him, even if only for the duration of the crisis. Another protracted negotiation ensued, during which the mayor proposed that an outsider take over the local police during the emergency, a dodge that Wright rejected out of hand. Once again the army officer felt compelled to beg for a resolution: "Give me any two-fisted man...no matter who he is." The city's highest-ranking official then settled on traffic lieutenant T. O. Miller. The officer proved a worthy choice, earning the military's praise for his deft and collaborative leadership.[18]

That should have been the end of the matter. After all, as city officials dithered, people were drowning. But Black would not let it rest. On Thursday, September 15, the day soldiers withdrew from the fast-recovering downtown business sector, he let slip to the *San Antonio Express* that Wright had "played politics" and used the army to force him to replace Mussey. The military counterattacked: in a series of private conferences, the same unnamed officer who had been involved in the earlier negotiations met with Wright and the mayor. Their interchanges apparently became increasingly heated, according to eavesdropping journalists. In the end, a truculent Mayor Black conceded that Wright had acted appropriately.[19]

This should have concluded the drama, but there was a final act. Once again the rivalrous Black and Wright were the leading players in the contretemps, with the army cast in a supporting role. The script centered on who should have had access to the business sector as soon as the floodwaters began to recede, and who had the authority to issue "passes" so that these select individuals could move in and out of the district. "When the first excitement of the flood was at its height," the two officials debated these related questions: their initial compromise was that they, plus a number of unidentified citizens, would have the authority to issue official passes to downtown businesspeople and property owners. Naturally, this only confused the situation on the streets. Because their ad hoc decision was not communicated to the military authorities, MPs and sentries found it "almost impossible to know whom to let through the lines and whom to stop."[20] The guards did their best with incomplete information, some leading citizens who were turned back came away with bruised egos, and the mayor and commissioner traded veiled and not-so-veiled attacks.

Those tensions had surfaced early on Sunday, September 11, as dawn broke over the still flooded city. Local politicians hastily assured military commanders that their armed presence was no longer needed. This knee-jerk reaction confounded those on the front lines. An army lieutenant later criticized the public officials for not knowing "what was happening down in the central district of the city where millions of dollars of property laid exposed to the hands of the looter." Once military officials pointed out the ongoing threat, the mayor and commissioner did an about-face, rescinding their order for the army to return to base. The battling politicians also clarified which of them could issue official passes: going forward, there would be one formal document issued containing Wright's signature.[21]

Coming under fire, too, was the army's gracious loan of thirty heavy-duty trucks to ferry supplies to Red Cross shelters and other distribution points, and to haul away tons of debris piled up in the streets, rivers, and creeks. This time, the combatants were parks commissioner Ray Lambert and street commissioner William O. Rieden. The latter complained to the *San Antonio Express* that Lambert had refused to grant his personnel use of the army vehicles, an allegation Lambert refuted by noting he had never been asked to lend them. Moreover, because Colonel Kromer, the military commander, had stipulated that the city would be liable for any damages to the trucks, Lambert thought it advisable to guard the equipment. "When men in Mr. Rieden's department approached the guards on Wednesday night with the request that the trucks be turned over to them, the guards at first refused, but upon learning who wanted them and for what purpose... they granted the request." Lambert aired this series of events in an open city council meeting, drawing "from Commissioner Rieden the admission that every request by him since the flood for trucks has been granted and that the two departments have worked in accord at all times." Having pummeled his overmatched opponent, Lambert then slammed the *Express*'s coverage to the delight of its competitor, the *San Antonio Light*: "The park commissioner severely condemned all attempts to stir up discord in the administration at a time when every resource... was needed to clean up and rebuild the city after the flood disaster."[22]

As petty as these debates appear, and as misguided as their timing was, the struggles between Mayor Black and Commissioners Wright, Lambert, and Rieden were woven into the DNA of the commission form of government. Originally the brainchild of Anglo elite reformers who wanted to replace the working-class machine that had controlled local politics since the late nineteenth century, the structure, which went into effect in 1914, included a mayor and four commissioners (parks and sanitation; police and fire; taxation; and streets and public improvements). Although this system was supposed to shift the balance of power to the handpicked representatives of the moneyed class, it was quickly co-opted by the politicians it was supposed to oust from office. These new commissioners just as speedily developed a keen appreciation for what they

must do to remain in city hall—provide tangible services to the citizenry, such as more parks; a larger presence of police and fire officers; and better streets. Ray Lambert, who had held public office since 1903, made certain as commissioner of parks that every bond issue included funds to acquire new open space across the community. Wright, who had risen through the ranks of the fire department to become its chief before his election as commissioner of police and fire, engineered bond funding for the construction of substations for his two departments. To win election, commissioners had to defend their turf; to hold their ground, sometimes they had to undercut their commission colleagues. When Wright and Lambert confronted the mayor and Rieden during the disaster, they were acting according to these rules of engagement.[23]

Once the takedown had been achieved, the winning commissioners would signal their triumph by seeming to paper over their differences with the vanquished. So, as soon as he flicked aside the challenge that Rieden had raised concerning the deployment of the army's trucks, Lambert quickly offered a resolution of appreciation to the army for the rescue and recovery work it had spearheaded during the crisis. His resolution gained unanimous approval, which must have been a relief to the military officers in attendance, given that their every effort had been met with some kind of resistance or ingratitude (or both). San Antonio politics, they had come to understand, was its own form of trench warfare.[24]

Lt. Dan Walsh Jr. was also struck by the baffling crossfire, and relatedly by what he perceived to be the city's fraught relations with those who served their country in its armed forces. This fickleness seemed to mock San Antonio's much-cherished nickname, the Mother-in-Law of the Army, a reference to the frequency with which officers, soldiers, and airmen stationed on the city's many military installations had met, courted, and married local women (among those meeting his future wife in the city was Gen. Dwight D. Eisenhower).[25] But Walsh was not feeling the love and voiced his critique of this on-again, off-again relationship in the *Trail,* a monthly magazine published by the army's Second Division that he edited from his office at Camp Travis.

In the twenty-one-page special issue—dubbed a "flood extra"[26]— and in prose purple enough to be worthy of William Randolph Hearst, Walsh gushed over the many emergency services army personnel had provided during the disaster. As the furious waters crashed into the city, he wrote, lives "were snuffed out in a twinkling of an eye. Hotels were piled up and business houses wrecked in the most horrible abandon. The injured cried for aid. The shivering huddled together in the ruined streets. The lights of the city were gone. The telephones were stilled. The police and the fire departments had exhausted the facilities at their command. Heroic efforts were needed. That was the Army's job." As soon as civilian authorities asked for the military's help, the Second Engineers, the First and Twentieth Infantry Regiments, and the Twelfth and Fifteenth Field Artillery rolled out. "Into the boiling waters the soldiers plunged" to rescue the drowning and to pull others out of their shattered homes; "during the dark hours of the night the glorious work of charity and valor continued." As day dawned, and over the next week, "they searched for the dead in the tangled wrecks and slimy creek bottoms. They cared for the sick and injured." As they guarded the devastated sectors against looters and marauders, they cared for the sick, injured, and hungry: "The food that was sent to feed and cheer the weary bodies of our men was gladly given the refugees," a commitment born of the same spirit "that three years ago steeled them to hit and break the supposedly invincible line of the Bosch [at] St. Mihiel" in France, and that in turn "inspired their souls to sacrifice in San Antonio because their job was just the duty of the day."[27] There could no question, the *Trail* asserted, that "the soldier was the salvation of San Antonio in its hour of need."[28]

Once that hour had passed, would the community remember those who had saved it? True, there had been an immediate outpouring of support, "a comradeship between soldiers on duty and the citizens with whom they came in contact." The publication singled out for praise those businessmen and "gentle ladies who missed no chance to get the sentry at his post a glass of cooling refreshment or to place their cars at the disposal of any man in uniform who needed 'a lift.'" In those moments, the camaraderie that had prevailed in San Antonio during World War I, "with patriotism at fever heat, seemed restored." Yet Walsh was also skeptical that this restoration of good feelings would prevail. "The army has been here so long that the people of San Antonio forget the history. Soldiers have always been on the city's streets. Pay-day has always been the city's biggest event. But—they were just soldiers."

The test of whether this more cynical detachment would resurface, he predicted, would come when next he and his comrades left their barracks and headed downtown to relax. What kind of reception would they receive when they queued up for an evening at the "picture shows"? How would locals respond when the "khaki-clad" sought new friends who would remind them "of the mother and the sisters they left behind when they joined the colors"? Should doors—real and metaphorical—be closed in their faces, if "nowhere remains open for their recreation but the joints that sell tequila or where their bodies may be scarred by the tattooer, then all has been in vain. We might as well quit and conclude that San Antonio, like most other cities, cares nothing for its Army except what it can get from it."[29]

An audacious *Trail* threw down the gauntlet, challenging city leaders, including "the Mayor, the Bishop [and] the President of the Chamber of Commerce," to give more than verbal assurances of the community's gratitude: "Actions speak louder than words." Almost in the same moment, and sensing that the newspaper might have overstepped, Walsh closed on a more upbeat note. For many soldiers, the disaster "was the first opportunity to have proof that the people are interested in their welfare." At the same time, the "army is as grateful to the good folks of the city for the kindness and appreciation they showed us, as the citizens are for the promptness and willingness of the soldier's response to their hour of need. This is our mutual opportunity to make this splendid relationship endure."[30]

Not surprisingly, the army's official report on its activities during the flood was much less personal and a great deal more tempered. It was also considerably less accessible. Unlike the emotional transparency of the *Trail*, and its public distribution—it was sold in local venues and portions of it were reprinted in the *San Antonio Light*—the same could not be said for the succinctly titled "Report on Flood, San Antonio, Texas, September 9–10, 1921." Compiled by the Engineer's Office of the Eighth Corps Area, and submitted to the Department of War, its tone is matter-of-fact and devoid of rhetorical position-taking. Striking, too, is that the document does not appear to have been shared with city officials, and there seem to be no public references to it.[31] Flying under the radar may have been what the US Army wanted. The community, however, would have learned much from this careful compendium, for it is the single most comprehensive account of the 1921 flood. It also reveals how World War I had enhanced the army's capacity to survey, document, and analyze landscape-scale destruction.

Long on detail, the report provides a minute-by-minute timeline of the army's actions, from the moment police and fire commissioner Phil Wright appealed for aid late in the evening of September 9 to the final pullback of all units six days later. Similarly meticulous is the documentation of the rotation of officers who were assigned to manage numerous command posts and their missions. The army's whirlwind of activity comes off as calibrated and controlled, which may also account for what is not included in the thirty-page, single-spaced, legal-sized report. There is, for example, no mention of the contentious political environment officers and troops marched into, which had complicated the military's operations, and no recitation of the verbal fireworks that erupted at checkpoints, command posts, and city hall. For the sake of harmonious relations, perhaps past difficulties and disagreements were best left in the past.

But this recapitulative text is also forward-looking, with a prospective point of view embedded in the report's data sets. These pull together all the known weather and rainfall data and chart the foot-by-foot rise and fall of the major rivers and creeks, as well as their varied velocity. The report offers a close analysis of the structural components of the city's streets and bridges and provides a thorough assessment of the damages they received. Taken together, this information lays a baseline for understanding where and why the flood occurred, a reference point for future inundations.

Every bit as intriguing is the material buried in the report's appendices, maps, and photographs, which reveals that the army's engagement with the flood went well beyond its intense rescue work. Its engineers and pilots mapped out the paths of devastation with the same precision that marked their reconnaissance and surveillance operations during World War I. Indeed, it is as if the disaster provided the military with the perfect opportunity to test its technical abilities and logistical capacities in a crisis. Military analysts, for example, tabulated the damage to the five railroads that crisscrossed San Antonio (a negligible $10,000); to municipal property, including streets, buildings, parks, and equipment (more than $400,000); to such critical utilities as water, gas, electricity, and streetcars (which amounted to an estimated $220,000); and to real estate, personal property, and merchandise (a more substantial $2.9 million).

This accounting also offered the army the opportunity to calculate what happens when disaster strikes, and at what cost, whether as a result of warfare or flood. These important fiscal lessons were tested by teams of military engineers who walked along the local waterways, ground-truthing the hydrology, calibrating streamflow, and developing cross-section profiles of channels and confluences. These data were represented visually in the blueprint-like documentation of the watersheds and photographs of crucial transit points, specifically every bridge that arched over the San Antonio River and San Pedro, Alazán, Apache, and Martínez Creeks. Generating this information was easier (and far less dangerous) than it had been during World War I, but the purpose was the same: to identify what had happened and why. There was an added bonus: the 1921 flood gave the army the opportunity to train new personnel in the art of post-battle damage assessment.

Army Air Service pilots stationed in San Antonio also tested their skill and equipment, and like the land-based engineers' evaluations, their overflights were as much about depicting how the flood tore through the built environment as they were about assessing technological advancements in airplanes and cameras that had occurred since the war. The outcome of the War to End All Wars had depended heavily on air power—not the much ballyhooed dogfights, but the airplane's role in providing timely and close air-to-ground support. Nothing was more important in this regard than aerophotography. The "aerial photograph," one contemporary French analyst declared, was "born of trench warfare," an insight that military historian Terrence Finnegan underscores. "Aerial reconnaissance and photographic interpretation reinvented the way that modern battle was envisioned, planned, and executed." The use of camera-mounted airplanes to photograph enemy entrenchments, artillery sites, ammunition dumps, and supply lines in real time offered the warring forces hitherto unavailable perspectives of the front lines and supportive infrastructure. "The advantages of an elevated platform capable of covering a broad area of enemy territory made photographic images an invaluable commodity for assessing enemy intentions." The value of this photographic evidence grew over the course of the war in tandem with the increased sophistication of cameras with ever-more proficient and multiple lenses that provided cleaner imagery and greater resolution. The airplane had become "the modern scout."[32]

In peacetime, there were fewer real-world opportunities to expand the range of the airplane's role as a reconnoitering technology. That probably accounts for the military's pilot-training forays into civil affairs between 1919 and 1939. In most cases, these interwar engagements were at the behest of other federal agencies or local officials, and invariably the military aviators flew supportive missions. One of these collaborative projects, writes historian Maurer Maurer, came at the request of the US Forest Service and the Los Angeles County Fire Department for the Air Service to conduct aerial surveys of the Santa Monica Mountains in Southern California in aid of wildland firefighters. In the southern cotton belt, army pilots tested pest- and crop-dusting in tandem with a US Department of Agriculture initiative. When the governor of West Virginia wanted to intimidate striking coal miners, he called in a compliant Army Air Service, which threatened to "bomb" union organizers with poison gas. More benign, in 1919 and again in 1922, pilots from Kelly Field flew in food and other supplies to a flooded Rio Grande Valley.[33] It is intriguing

that Maurer did not know about, or at least did not comment on, the surveillance that a pair of Kelly Field–based aviators conducted within days
of the 1921 flood. An almost cryptic single sentence in the army's flood
report signals their mission: "1st Lts. Halsey L. Bingham and Harold R.
Rivers, Air Service, were assigned the duty of taking aerial photographs
of the flooded district."[34]

Their visual record would bring the catastrophe to life because their
overflights tracked the flood's movement down three key watersheds.
Flight 1 surveyed the San Antonio River and its serpentine course from
just south of Brackenridge Park through the downtown core and then
followed the river's path to South Alamo Street and its environs. Flight
2 covered a critical portion of San Pedro Creek—from the Union Stock
Yards, just south of that creek's confluence with Apache Creek, north to
the confluence of Alazán and Apache Creeks. Flight 3 flew north and west
from the stockyards, following Apache Creek to West End Lake (now
Woodlawn).

A century after the shutters snapped on these scenes and army interpreters had pieced together the resulting black-and-white photographs
into a seamless whole, the stark mosaics still offer chilling evidence of
the flood's extent and its ground-clearing force. It is easy to spot the San
Antonio River's damage as it powered out of its banks and raced south
and west. As readily seen is the path of the rampaging water as it tumbled
down the San Pedro, Alazán, and Apache Creeks. The color tone offers
one clue: streets and lots that are a brownish-white were badly damaged
or swept clear. Here homes, buildings, and vegetation were sucked into
the flood's whirlpool-like energy and shot downstream. Streetscapes with
a darker hue appear to have escaped the heaviest damage, and military
interpreters rendered these distinctions more visible by laying down a
black-ink line indicating the outer edges of the most seriously flooded
areas.

Inside those demarcations, on the West Side especially, lay a landscape of death. Not that the army called out the dramatic links between
who died, where, and why, but the photographic evidence makes those
connections painfully clear. What was not obvious to the army's engi-

neers was what kind of structures might prevent another such tragedy in the city's poorest neighborhoods. "Flood prevention on the San Pedro, Alazán, Apache and Martínez Creeks presents certain difficulties not present in the study of the San Antonio River," the report acknowledged. Perhaps the most important and underreported of these difficulties was that "no one seems to have foreseen that as much water was liable to have come down these streams as down the San Antonio River." Mitigating their storm-driven flows and velocities, however, was complicated by this sector's topography. Unlike with the San Antonio River's watershed, the report indicated that "no satisfactory site is believed to exist for an impounding reservoir of sufficient size" on the West Side. Another compounding factor was that the "capacity of the existing channels of these creeks, in comparison to the flood that they were called upon to carry, is so small that no improvement in the channels themselves would be sufficient to provide a waterway for these floods."[35]

What was possible was determined in part by the West Side residents' poverty and the sector's poor housing stock: "In view of the small property values and less built up condition, certain other things, not feasible on the River, can be considered practicable in these creeks." Among the army's proposals was the prohibition of post-flood development close to the creeks so as to maintain open strips of land along their banks, "the width of which would vary according to the topography." The army also proposed that creek channels should be excavated where possible and the "excess dirt taken out in widening and deepening these channels may be thrown up on the edge of the reserved strips in small levees, or may be used, in some cases, to fill up the low ground."[36]

The goal of creating a wider floodplain along the West Side creeks was more benign than another strategy that the military and the local flood prevention committee had entertained. An early proposal had been to build upstream diversion dams in the Olmos Creek watershed to redirect its floodwaters down the Alazán. In remarks to the local Lions Club, county judge Augustus McCloskey, a significant figure in San Antonio politics and an active if ex-officio member of the flood prevention committee, commented that "until some means are taken to divert the flood

Multiple railroad tracks cut through the West Side, 1939.

waters of the Olmos valley into channels other than the San Antonio river," the floods that "had visited San Antonio for 200 years will continue to do so."[37] Ten days later his remarks gained added credence when a board of engineers working as a subcommittee of the flood prevention body released an outline of its working suggestions for flood control projects, including diversion of Olmos Creek's floodwaters to the Alazán.

This possibility suggests that the engineers understood that the West Side already was a sacrifice zone. Multiple railroads sliced through it; several offal-smelling stockyards occupied large swaths, along with lumber mills, laundry facilities, ginning plants, light industry, and bars, saloons, gambling emporia, and houses of prostitution. While flood control planners were debating whether to channel more stormwater down the Alazán, city commissioners were adding to that section's environmental and public health burdens by authorizing the construction of a forty-ton-capacity incinerator at West Travis and North Las Moras Streets, one block west of Alazán Creek.[38] These were signals to engineers that it was acceptable to seriously consider disposing of diverted floodwaters there.

Other indicators led them not to pursue a similar diversion for Salado Creek, which loops around the city to the east. C. H. Kearney, former city engineer and a strong proponent of what would become the Olmos Dam, dismissed the Salado option within a day of the flood: it would be an expensive proposition, he told the *Express*, not least of all because "it would involve the city in a never-ending lot of damage suits from property owners along the creek."[39] White ranchers and farmers had the fiscal and legal resources to protest and litigate, so their interests were deferred to in advance. Poor renters on the West Side did not, leading others to arrogate to themselves the authority to determine their fate. The water engineers, for example, recognized that a diversion of flood-high flows would have rendered much of the West Side uninhabitable and noted that for the scheme to succeed the West Side creeks would require "additional protection." Fortunately, once they ran the numbers, the board abandoned the Alazán plan as too expensive.[40] Yet the idea did not disappear. Thirty years later, as the US Army Corps of Engineers evaluated the potential for new flood control infrastructure in San Antonio, it reported to Congress that it had considered a diversion scenario for some West Side creeks before deciding that this did not pencil out. That the concept had persisted, that the West Side continued to be thought of as expendable, is an indictment of those who had the power to plan its future.[41]

As it was, creek widening was a considerably cheaper and much less sophisticated form of flood control than what San Antonio's power elite planned for their prime area of interest and concern. What the chamber of commerce demanded, what engineers, newspaper editors, and every public official threw their support behind, was the construction of the $1.6 million Olmos Dam. By cheering the "vigor" and speed with which the chamber and the flood prevention committee had formulated plans to "prevent recurrence of the disaster in the future," the army allied itself with this powerful coalition's conviction that "protection of the business district is vital to the city." A final series of aerial mosaics are a product of this alliance. While aviators Bingham and Rivers flew over the San

Army aerial mosaic of potential cutoffs to speed floodwaters through the central core.

Antonio River, they zeroed in on those sections that the 1920 Metcalf and Eddy report had stipulated should be shortened and straightened. "Most of the cut-offs recommended by Metcalf and Eddy expedite the passage of water down the congested business section," the army's report confirmed. "It is equally as important to carry water from this business district quickly and also from the residence district below." The new channels would prevent floodwaters from jumping the banks at tight turns, such as routinely occurred at the Great Bend: "By eliminating the Great Bend, a great length of the river will be avoided where the closely built up section makes any large change in the cross section of the [current] channel most difficult." The military's photographic interpreters provided a graphic representation of these observations by overlaying aerial images of the river with a set of dashed lines to signal where the cutoffs would be constructed.

This visual record, combined with the military engineers' technological orientation, reinforced the city leadership's fixation on the river's threats to downtown economic interests. The military became complicit

in the Anglo elite's small-group decision-making process that largely ignored other options and voices, and that further marginalized the city's
most impoverished residents. The end result, as every subsequent flood
demonstrated, was neither neutral in its application nor abstract in its
consequences. Persisting for the next fifty years, this set of social inequities and environmental injustices would continue to grieve those calling
the West Side home.

-[FOUR]-

DAM THE OLMOS!

Floods are episodic and ephemeral. Flood control infrastructure is enduring and perpetual. The latter is designed to prevent the former. Certainly, that is what a dam's builders hope their project's legacy will be. It is no surprise, then, that protection from raging floodwaters and the collective peace of mind that would bring was uppermost in the minds of those selected to speak at the public dedication of the Olmos Dam on Saturday, December 11, 1926.

The central figure at that celebratory event was San Antonio mayor John W. Tobin, and Judge Arthur W. Seeligson, who served on the city's flood prevention committee and who delivered the principal address, made certain that the audience understood Tobin's centrality. In a mere three years in office, the mayor had engineered a series of successful bond elections that built sewers, parks, and fire and police department substations; hardened streets; and underwrote the construction of grand public buildings like Municipal Auditorium. But the most crucial of these initiatives, Seeligson affirmed, was the $2.2 million bond that paid for the construction of the flood retention dam on which the assembled throng stood. By his concerted actions, "Mayor Tobin has proved himself to be one of the greatest civic leaders in the history of San Antonio."[1]

An adept politician, Tobin deflected the (probably justifiable) praise by framing his remarks around the two-thousand-foot span that now sealed off the Olmos valley, once the source of the major floods that had scoured the Olmos Basin and San Antonio valley for centuries. He ap-

Olmos Drive facing west over the dam, toward Olmos Park.

plauded the many citizens who had voted for the bonds underwriting the dam's construction—it was their "splendid cooperation" that had made the project possible. He roundly cheered "flood prevention engineer" Col. S. F. Crecelius "for the part he played in erecting the dam" and A. J. McKenzie, the project contractor. Those niceties completed, Tobin told his listeners that what really mattered about the gleaming structure, and what justified the $1.6 million that had been poured into its construction, was that it provided unprecedented security. "San Antonio's old flood nuisance is now ended," Tobin assured the audience, "and there is nothing to stand in the way of her progress and growth." The city's future beckoned bold and bright: the dam's construction was "the greatest step that San Antonio has taken towards becoming the leading city of the Southwest."[2]

Yet perhaps the most striking element of the dedication ceremony was not the orators' rhetorical flourishes and boosterish claims, but the number of automobiles at the event. Earlier that morning more than a hundred vehicles had rallied at Municipal Auditorium, itself recently erected on a triangular site created two years earlier when the city had straightened a double bend in the San Antonio River's serpentine course. From there, the lengthy convoy carried public officials and residents north up McCullough Avenue to the brand-new two-lane Olmos Drive. They turned right and headed east toward and then onto the dam. The parade pulled to a stop halfway along the span, where it was met by another fleet of cars ferrying residents of Alamo Heights, an incorporated suburb abutting the dam on its eastern edge. In the lead car was the small community's mayor pro tem, W. H. Hume. The two public officials, Tobin and Hume, stepped out of their respective vehicles, walked to the center of the causeway, and shook hands. With cameras clicking, they exchanged pleasantries. Tobin offered that it was "nice to have Alamo Heights as a neighbor." A starstruck Hume replied: "And it is worthwhile to be the little brother to San Antonio and have beauty spots tendered to us like you have in building your dam and establishing your wonderful park here."[3]

Beneath the benign banter lies an important message about the reciprocal relationship between the internal combustion machines parked on the dam, the tons of concrete that gave the dam its form, and the larger flood control project for which the imposing structure was the linchpin. All were emblems of a modernizing San Antonio, harbingers of its grand future. The automobile—and the enhanced mobility it provided to those with means—would power a physical realignment of the city and its emerging suburbs. Hereafter, too, the city would be cast in concrete, a material applauded for its effectiveness, efficiency, and aesthetics; the material's sunny-day gleam caught the imagination of the city's flood prevention committee, who chose it over building a more drab earthen structure.[4] As for the dam itself, it would only protect lives and property within the San Antonio River's floodplain to the south, not in other flood-prone sectors of the community. These significant limits to the dam's protective impact did not deter Mayor Tobin in 1924 from demanding its prompt construction. Any dallying would lead to inaction, he had asserted in August of that year, and "the city had already delayed too long in carrying out the flood prevention program."[5]

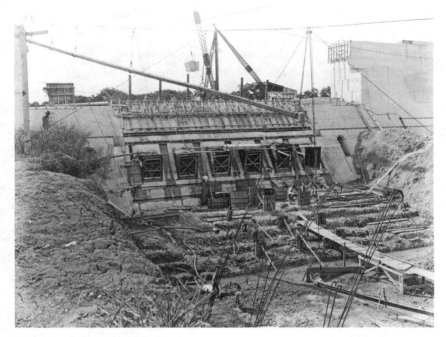

Olmos Dam under construction, August 1926.

That the Olmos Dam was built at all is something of a miracle. After all, floods had repeatedly ripped through the region dating back to the city's founding in the early eighteenth century. The record of these earlier inundations is sparse, so it is not clear what the reactions to the floods were, but it is reasonable to assume that they were in line with residents' responses following the well-documented flood of 1819. When devastating floodwaters roared through the adobe-built town, its New Spain leadership, both secular and sacred, did not seem to consider how vulnerable it was. Or, rather, because they were utterly dependent on the well-engineered system of acequias, or irrigation ditches, and the river for potable water and irrigation, they may have felt they had no choice but to rebuild what that flood had torn asunder.

Subsequent generations at least paused before reaching the same conclusion: in most cases, as soon as the swirling waters had receded and the dead had been buried, public officials and residents vowed to buttress their precarious existence in the San Antonio River valley. They did not pause for long, though. The first of these partial reckonings may have been in the aftermath of the 1845 flood. The stormwater only intensified

communal anxieties over the ongoing uncertainties of living on the edge of the new state of Texas, with its tensions with Mexico. "It caused considerable damage to the city and played havoc with the houses near the banks of the river," the *San Antonio Express* reported, noting a well-known and destructive pattern; when confronted with a curve or bend, fast-moving waters would race out of the river channel and slam into houses and buildings. Recognizing how vulnerable the community continued to be, the ayuntamiento, or city council, announced that it would "move the pueblo to higher altitude," relocating it to Franklin Square just west of San Pedro Creek.[6]

This ambitious effort to pull the city center away from the river met with a storm of protest. "Common opinion did not favor the City Council's opinion because the new site was too far from the church and the old pueblo. So public sentiment prevailed, and the people had their own way in the matter." Parsimonious Anglos were no more interested in taxing themselves to straighten the San Antonio River than they were in building a dam to impound Olmos Creek—two strategies that mayors Charles F. King (1846–47, 1853) and Sam Smith (1847–49) repeatedly championed "as the only effective means of protection against the ever-impending danger of flood." The citizenry apparently preferred to be stuck in the mud.[7]

They were no more willing to pull themselves out of danger in the decades that followed. In 1852 another flood tore through the city. Its flow and damage were less significant than during the 1819 disaster (at its peak, the river in 1852 was measured at 4,000 cubic feet per second (cfs), or roughly 30,000 gallons a second; by contrast, a retrospective analysis of the 1819 flood estimated its surge at more than 30,000 cfs, or 225,000 gallons a second). Nevertheless the overflow was powerful enough to hammer bridges, undercut buildings, and disrupt daily life. Yet when local engineers and political leaders raised the possibility of underwriting flood prevention measures such as those proposed in the 1840s, the community again demurred.[8]

Thirteen years later, a more powerful flood rampaged through San Antonio with a flow of 7,000 cfs, nearly double that of the 1852 event. Accompanied by a fierce hailstorm, the river rose nearly eighteen feet, according to one report, and swept away livestock, horses, mules, and chickens. Once again, lives were lost, bridges and other infrastructure were damaged, houses and offices took on water, and streets were impassable for days. And once again, there were calls for an orchestrated effort to erect flood control measures. Most notable was a report written by three important figures in the city—Francois Giraud, an architect, surveyor, and future mayor; Gustav Schleicher, a member of Congress; and Victor Prosper Considérant, a former city engineer. "We are aware of the deep feeling produced in this community by these late disasters," they noted in the document they submitted to the city council. "Many families have lost their homes; merchants have lost their goods; properties of all descriptions have been lost or injured, even in localities always considered safe heretofore; and last and saddest human life has been destroyed."[9]

Although the trio of authors acknowledged that the "late freshet has been an extraordinary one, and it is beyond human power to avoid all the effects of such rises," they suggested immediate steps that should be taken to reduce the risks manifestly evident to all who lived in San Antonio. Among them was the clearing of all overgrown vegetation—trees and shrubs—that choked the channels and "forced water out of its bed." They also urged people to refrain from using waterways as landfills, tossing garbage and other debris into the river. During a flood, the abandoned material functioned much as invasive willows did: "Every obstacle placed in the river, so as to interfere with the free, clear passage of the water, has the tendency to add to the rise of the water in a freshet."[10]

Those practical actions came paired with, indeed required, an awareness of San Antonio's flood-swept past. It has become "a serious question whether the effects of these floods have not been increased by a want of foresight and causes which might have been avoided, and, if so, it is the imperative duty of the community to remove any such existing causes, and for the future, by taking all such precautions such as long experience has dictated in other jurisdictions similarly situated to prevent their rep-

etition." The reality of living in a city built into a floodplain demanded
no less.[11]

That reality led to an August 1868 public meeting in which downtown
merchants and property owners pressed the city to divert Olmos Creek
"into the Alazan to prevent overflows in the San Antonio River." This
may have been the first time that the idea of sacrificing the West Side to
benefit the commercial core was raised, but it did not gain traction.[12] The
consequences of the community's inaction, its willingness to forget history, became evident in early July 1869. A punishing storm pounded the
region with fourteen inches of rain and hail, the rivers and creeks burst
their banks, and an estimated eighteen houses constructed of limestone
or adobe were destroyed. The Nat Lewis mill at the Navarro Street crossing and the adjacent pedestrian bridge would have been heavily damaged
if the quick-thinking mayor, Wilhelm Thielepape, had not come upon
the scene. A massive tree trunk racing down the swollen river "became
jammed between the wheel of Lewis' mill and the foot bridge, and the
force of the stream was gradually pushing the bridge from the abutments,"
the *San Antonio Daily Express* recounted. Alerted to the dangerous situation,
Mayor Thielepape, in collaboration with several workers, "extricated the
tree and put the bridge in place."[13]

While the newspaper cheered the mayor's "prompt action," it also
recognized that such ad hoc efforts were beside the point. In a lengthy
editorial, "What Would Relieve Our River," the *Daily Express* decried,
as had the authors of the 1865 report, the community's continuing inattention to San Antonio's flood-prone nature. "It requires no scientific
observation to know that if every obstruction in the way of dams, walls,
rubbish, trees, etc., were removed out of the river from the city to its head,
the water would run off more rapidly and at least three feet of rise could
be guarded against by this cleaning out, and three feet would put us on
the safe side."[14]

That sense of safety was contingent on the size of the flood. The
newspaper's writers did not know the San Antonio River and its larger
watershed could carry more than four times the flow that had raced
downstream in 1852. But the journalists recognized other benefits from

rigorously managing the riverbed: there is "no doubt but the former purity and brilliant clearness of the river water would be restored, and additional security added to the health of our city." This was a point of considerable merit given that San Antonio experienced regular outbreaks of waterborne illnesses such as cholera, typhoid, and dysentery (and the latter two erupted in the fetid wake of the 1869 flood). Meritorious, too, the newspaper argued, would be to extend this river cleanup campaign to the acequias. Once they were also "thoroughly relieved of rubbish," they would be better able "to carry off more water" and thus provide "safety values." Moreover, if the irrigation system was extended along the "east side of the river to skirt the foot of the hills and be carried far below the city," it would create added defense against overflows while expanding the agricultural economy. "Let us see if we cannot kill more than two birds with one stone, instead of throwing \$20,000 into a dry creek," the news story stated.[15]

This may have been the first proposal that explicitly linked flood control investment with downstream economic growth. It would take another sixty years before the linkage resurfaced in political discourse. This time, though, parsimony won out over protection. Late-nineteenth-century San Antonio taxpayers were notoriously tight-fisted, as unwilling to underwrite sewers and sidewalks as they were to use public money to clean the riverbed. Not for them the taking on of debt in the present to ensure their lives and livelihoods in the future. Just to be safe, though, well-off San Antonians, whose disposable incomes allowed them to move to higher ground, did exactly that and began commuting on streetcars between hilltop and floodplain. Their gain in elevation meant that henceforth the poor would bear the burden disproportionately every time floodwaters churned down river and creek.[16]

Such spatial inequities fed into a complacency that characterized much of the ensuing public discussion over the threat of flooding. A classic example appeared in a lengthy 1887 retrospective on how much San Antonio had changed over the previous decade. In it, the *San Antonio Daily Express* uncoupled the city's present from its flood-wracked past. "Destruction by inundation was the great bugbear of the inhabitants of San Antonio" in 1877, it recalled. Residents worried that Olmos Creek, "which becomes a torrent during heavy rains," would do "considerable damage to property along the banks of the river." They had reason to be anxious, the newspaper allowed. Citizens would "sit in their offices ten years ago and figure out on their maps the certain destruction which was to come. No amount of argument could get them out of their idea. The Olmos would surely destroy us. They were certain of it. It was only a question of time. It was bound to come." The *Express's* use of the past tense, a distancing device as effective as the gently mocking tone, implied that such worries had been allayed. "Yet we are all here, and every possible danger from the Olmos has been averted merely by converting the hard, unbroken prairie into cultivated fields. The man who ventured to predict an overflow today would be called an idiot."[17]

Two years later the *San Antonio Daily Light* contributed to what might be dubbed a conspiracy of optimism. On July 5, 1889, the afternoon newspaper marked the seventieth anniversary of the 1819 flood with a short, ten-line article, "San Antonio's Johnstown," a reference to the killer flood that six weeks earlier had killed more than two thousand in Johnstown, Pennsylvania. The article might have used that tragedy to warn that a similar fate might befall San Antonio, as testified by the death and damage the 1819 flood had caused. While the text does not make light of what had happened in San Antonio, it does not function as a cautionary tale: "Seventy years ago today, a cloudburst north of this city raised the river and San Pedro Creek so high that their waters met on Military Plaza. Histories in the possession of County Clerk Smith say over 100 houses were washed away and several lives lost." Like the documents in the archive, that troubling past had been buried.

By 1903 the historic 1819 flood was lost to memory. Or so it would appear from the *Daily Light's* effort to place the February 27 flood of that year in its proper context. Even as the paper editorialized that "the rain gauge tells that there is not enough rain falling in this section to flood the country. Not the least need of getting scared," it acknowledged elsewhere in the same issue that this particular inundation was nevertheless "too high for comfort." That said, the current flood did not measure up to some of its predecessors: "And the rains descended and the floods came and the San Antonio river was higher than it had been in a third of a century, the highest water that transcended that of Thursday morning having been recorded in 1870. The rise of 1865 is the highest known." (The *Express* reaffirmed this claim after the October 1913 deluge, when it anointed the 1869 event "the greatest flood in the history of San Antonio outside of the one today.") For early-twentieth-century San Antonians, it was as if the 1819 flood had never happened.[18] —{ 107 }—

Like several splashes of freezing cold water, it took three major floods—two in 1913 and one in 1914—to shock the city out of its culture of complacency. The first of these floods crashed through the city on October 1, 1913. A powerful storm that dropped seven inches in a day sent water roaring down the West Side creeks and the Olmos Creek / San Antonio River basin. A purple-prose headline in the *San Antonio Express* captured the storm's devastating energy: "Fed by incessant rains limpid stream becomes raging torrent." It was a roar of water, the article said, that had silently "crept into the city. In its race to the sea the volume each minute became greater and rose higher until the narrow channel of the river burst, spilling the flood over the lowlands and into the houses of hundreds who slept unconscious of their danger. Not until the drift-filled current began battering houses and the chilling tide rose waist deep in their bed chambers did many awake." There was considerable damage "in the flats of the west side drained by San Pedro, Alazán, and Apache creeks," noted the *Light*. "A great many of the frame box-houses have been floated off, turned over, or moved from their foundations," damage that when coupled with the loss of food, clothing, and furniture consequently fell on "those least able to sustain it." A mother and her three children drowned attempting to cross the San Antonio River near Mission San José.[19]

Although the *Express* argued that the flood was "almost unprecedented" in its rise, that was not the case. It was measured at an estimated 7,200 cfs, a figure the 1819 flood dwarfed fourfold. The 1913 flood was much more like the modest—though still dangerous—floods of the mid- to late nineteenth century. The key difference was that by 1913 San Antonio

Whitecaps in the October 1913 floodwaters.

was home to more than a hundred thousand inhabitants, five times more than its population in 1880. These greater numbers of people occupied more space and lived in greater density than their predecessors. When the water rocketed into the river's serpentine stretch north of downtown, rose out of its banks, and streamed down Saint Mary's Street and a good many others, and when the surge bottled up in the West Side creeks and overflowed, the waters swamped a more expansive street grid and inflicted much greater damage. Hotel basements became catchment basins, cellars became pools, bridges cracked, and those buildings and houses propped up on weak foundations collapsed or broke free. It is no wonder that police and fire units, as well as troops from Fort Sam Houston, found so many "men, women, and children cut off by swirling and muddy waters, huddled alike in humble cottages and more pretentious homes."[20]

Two months later, the same scenario: Water from a thunderous downpour cascaded down the region's watersheds. Olmos Creek rose fast and collided with the San Antonio River, forcing floodwaters to barrel down River Avenue, Saint Mary's Street, and other north–south conduits. Once more, first responders plucked people from the river, pulled others out of windows and off roofs, and snagged those clinging to trees. It all seemed routine. Yet the December inundation—with an estimated flow of 8,000 cfs—gave a shock to the system. No one could remember when two floods had scoured the city in such short order: "After the flood of October, even the old timers said, 'This will not occur for twenty years, yet within 62 days a greater flood swirls through the city.'" Neither reporters nor the hundreds of sightseers from high-and-dry parts of town who came to splash in the flooded streets or to gawk at the rushing energy coursing beneath the viaducts could recall a time when such a seemingly insignificant amount of precipitation had generated such a "spectacular" rise in the river and creeks. Never before, it seemed, had every bridge in the county been "damaged or swept away," a reality that left "the highways running out from San Antonio in anything but satisfactory condition" and placed the regional economy in a precarious position, however short-term the hit.[21]

The city's economic activity would recover—shelves would be restocked, housing repaired, hotels and other commercial buildings pumped dry and touched up—just in time for a third "freshet," this one almost exactly a year after the first. On October 23, 1914, a slowly moving thunderstorm dropped five inches of rain over the West Side, whose "usually tractable and harmless little streams proved wholly inadequate to carry off torrential waters falling at an inch an hour and rushing into them from more or less precipitous watersheds that quickly rocketed down the narrow creeks."[22] The Alazán, a "seething torrent," proved the most dangerous. Much as in 1921, the turbulent creek ripped through low-lying neighborhoods and splintered the shacks in the "corral" district. As it sucked up debris, it knocked people off their feet, ripped out trees, and bludgeoned bridges, railroad trestles, and other infrastructure. "It is conservatively estimated that 2,000 people have been rendered homeless and all but destitute, either by the total destruction or the inundation of their dwelling places. A very large number of the very poor class lost practically all their goods and chattels."[23] The rampage claimed nine lives, the majority of whom were members of the multigenerational Liebe family (the youngest, not yet named, had been born a few hours before the stormwater flattened the family's home). "Only one person is reported as missing, an unidentified Mexican boy," the *Express* noted. "The boy was last seen on the top of a door floating down the stream."[24]

That haunting image, paired with those of helpless mothers and children yanked from their loved ones, compelled the community, after decades of denial, to ramp up discussions about implementing significant flood control protections. One local daily was already on record arguing for concerted action. After surveying the wreckage following the October 1913 inundation, the *Express* delivered a series of warnings. It began with a lamentation: the city had been lulled by the fact that the river in the previous decade "had been so low that it was unable to get out of its bed and some persons believed that it would never rise again." Listen to the "oldest inhabitants," the *Express* urged. "They remembered what had occurred aforetime, knew better, and were not a whit surprised when the sluggish stream became a torrent that threatened dire destruction." Nature alone did not cause the extensive damage. The river's "domain had been too much encroached upon by the greed for building space," structures that served as walls in the historic floodplain and thereby elevated and accelerated the movement of water through the city. The solution was manifold. Clear the obstacles in the river and creeks, straighten the twisted contours of the river, but most importantly shut down Olmos Creek, which time and again had poured water into the San Antonio River that then slammed into the downtown core. "Suppose a large part of the Olmos' contribution had been impounded," the *Express* wondered, held back by an "immense reservoir above the head of the river, being allowed to come down only by degrees, after the reservoir was filled, would there have been any danger of overflow?" The answer was emphatic: "Dam the Olmos? Dam the Olmos!"[25]

To which the *Light* countered: "No Dam above the City." The newspaper's editors were all in favor of widening, deepening, and walling the river, but they were (gleefully) shocked that their rival would advocate for an upstream dam. San Antonio should never "for one moment consider the construction of a dam near the head of the San Antonio River or at any other point which will place the city at the mercy of the flood waters of such a reservoir. The idea is positively the most dangerous that can be suggested." To hammer home their point, they compared what might happen in San Antonio to what had occurred to Johnstown, Pennsylvania, when a dam collapsed in 1889 and wiped out that mining town. Given that the south-central Texas city "was much larger than Johnstown…the chances of disaster are correspondingly greater." So great that any such disaster "might easily dwarf all the damage San Antonio has sustained, or ever will sustain from the river as it is."[26]

Theirs was the lone, anxious cavil. By contrast, the chamber of commerce promoted saving businesses through some form of intervention like a dam, and political leaders appeared on board with that strategy. Asked if he was in favor of a dam, Mayor Clinton L. Brown (1913–17) assured reporters that he was, or at the very least he was in favor of straightening the river, dredging the waterways, and setting back buildings that crimped the river's course. When pushed about the possibility of floating a bond to pay for reconstructing the river and streetscape, and constructing a dam spanning the Olmos valley, he began to equivocate. "I haven't fully investigated the details of the Olmos plan," he declared, "but I am convinced there is merit to the suggestion." The key hindrance to moving the plan forward was its anticipated cost. "I have no idea what the cost would be, but if it is not prohibitive, its benefits would more than repay the costs of the investment."[27]

That was a large "if." As with his predecessors, Brown had had enough experience with local taxpayers to know that they were a fickle lot: they would be delighted to benefit from improved services—but not on their dime. No one was surprised, then, when over the next year the mayor and council did little to advance the cause of the dam's construction, inaction that reporters raised with Brown after the fatal October 1914 flood. Calling it "an act of God," he then asserted, all evidence to the contrary, that this last of the back-to-back-to-back floods was "without precedent" and "could not be foreseen" and insisted that under the conditions the damage could not have been prevented. Conceding that rebuilding the urban environment and the river and creeks that flowed through it "presents a difficult engineering problem," he underscored that paying for this ambitious river project represented an even "more difficult financial problem."

Brown revealed just how complex the politics of that problem were in the very next sentence: "I cannot say now, though, just what I will recommend to the Council."[28] When he left office in 1917, the question of the dam, and flood control more generally, remained unanswered.

For the media, the answer was clear. In a strikingly broadminded October 1914 editorial, the *Express* urged the citizens to demand change: "Time Has Come When the People Should Speak." Yet when they asserted their voice, the newspaper stated, they must do so in unison. "There should not be a spirit of selfishness manifested in any manner regarding the betterment of the river conditions." Chastened by the loss of life along the Alazán, the newspaper argued forcefully that the "straightening of the banks and the dredging of new channels to better the flood conditions in one portion of the city at the dreadful expense of death and damage to another part should not be considered." Instead only "the best interests of the whole interest of San Antonio's citizenship" should determine the array of flood control measures the city would undertake. "It is not what will protect this or that building or block but what will relieve the entire city of future fear and future toll of damage and death."[29] This inclusive vision was as principled as it was unusual in a city riven by longstanding class, racial, and ethnic divisions. It marks high ground in the local dialogue about the social, economic, and political benefits that flood control could engender for the entire community.

Over the next decade, however, this view would lose out to the much narrower interests of the city's commercial and political elite. Engineers aided and abetted this shift in focus. Science became the handmaid of discrimination. The key document in this regard is the 348-page report that Metcalf and Eddy, a Boston-based firm of consulting engineers, submitted to the city council in December 1920. Commissioner Ray Lambert had contacted the firm that summer because Texas-born Metcalf was well known to the city, having completed several projects for the San Antonio Water Works Company. To develop as complete a history of flooding in the region as possible, Metcalf and a colleague spent the next five months rooting through local archives, newspaper records, and city documents. They interviewed residents and public officials and combed through US Geological Survey (now Service) reports and climate data. They also tested what they had read against the lay of the land, walking up and down five major watersheds, including the San Antonio River and San Pedro, Alazán, Apache, and Martínez Creeks. By ground-truth-ing their historical research—much as the US Army engineers would do after the 1921 flood—the Metcalf and Eddy team members were able to set the record as straight as possible. Not only did they remind the city of what it had forgotten—that the flood of 1819 was significantly larger than any subsequent inundation—but they made its dangerous high-water mark the baseline for future flood control initiatives. In turn, this revision framed, contextualized, and prioritized the firm's catalogue of recommendations.[30]

In an important sense, there was nothing particularly new about Met-calf and Eddy's prescriptions. As a self-described "river crank" wrote the *Light* in September 1921, any "person with eyes can see that our town is in a valley which might be flooded out at any time there is a waterspout or very unusual rainfall in the Olmos watershed." That pragmatic per-spective had been repeatedly affirmed: "Engineers have been paid time and time again to tell us what any man with a little observation and common sense already knew."[31] The correspondent was right: the situa-tion and its resolutions were obvious. But the Boston engineers offered a full-scale reconstruction of the watershed, recognizing that piecemeal changes would never be enough to resolve San Antonio's historic flood-ing problems.

They started with the dam. As proposed, it would be sited just north of Brackenridge Park where Olmos Creek runs through a narrow valley, a pinch point, so that the structure would be slotted into and take advan-tage of the limestone bluffs on the creek's eastern and western banks. Its purpose was to impound floodwaters only; on normal-flow days, gates at the foot of the structure would let the creek continue to run as usual. This keystone infrastructure in the plan was essential to ensuring that San Antonio could withstand a hundred-year flood such as had ravaged the city in 1819 (and would again in 1921). The Boston engineers knew enough about the city's bargain-basement approach to public spending, however, to recognize that the structure's price tag, which they estimated at $950,000, might mean its construction would be delayed.[32]

They made that delay palatable by recalibrating the dam's contribution to flood control as a whole. Even though the detention dam alone would be of inestimable benefit, Metcalf and Eddy reasoned that "an extreme flood might produce a runoff from the drainage area below the detention basin which would be far beyond the capacity of the present channel." Put differently: "A detention basin alone would not completely control the floods of the San Antonio River and of the two methods, the improved channel alone would accomplish more than the detention basin alone." The engineers, like the city council, realized that deferring dam construction was a high-stakes gamble.[33]

Metcalf and Eddy's confidence in improvements in the river's channel downstream depended on a suite of alterations to its course. With surgical precision, they proposed six major cuts to the river's bewildering, tortured path. To the north of the densely developed central core, they envisioned slicing through a loop in the river just below Josephine Street, where the 1913 flood and earlier ones had shot out of the channel and raced down nearby streets. Several blocks to its south, they suggested a cutoff through a troublesome, convoluted bend between Eighth and Tenth Streets; for a tight loop at Trenton Street; at Romano Street (at the subsequent location of Municipal Auditorium); and at dangerous loops south of downtown, one at Martínez (now Durango) Street and another at the Guenther Lower Mill. These incisions would enable the river to carry roughly 12,000 cfs at flood stage (as a comparison, the December 1913 flood was measured at 8,000 cfs) and would shorten the river's course by 1.3 miles. During subsequent storms, water could move through and out of the city at a much faster clip.[34]

Metcalf and Eddy's final set of recommendations involved raising, strengthening, and/or rebuilding city-owned bridges. River-crossing railroad trestles "should be made to conform to a type that will permit the undisturbed flow of water." The river's bed should be cleared of vegetation and its banks stripped of trees, a recommendation that later drew sharp and successful protests. Less controversial were the gen-

SIX RECOMMENDED CUTOFFS TO THE SAN ANTONIO RIVER, 1920

Location	Length of river shortened (feet)
Josephine St.	1,200
Ninth St.	2,450
McCullough Ave.	950
Navarro St. / Romana St.	1,495
Martínez St. / Durango St.	670
Guenther Lower Mill	285
TOTAL	7,050

eral design specifications, which included excavating the river channel to twenty feet deep and seventy feet wide with reinforced concrete sides. Human-constructed obstacles, such as check dams for flour mills on the northern and southern reaches of the river, should be demolished. As for the major West Side creeks, the report was much less detailed than it had been for the San Antonio River. The single intervention would be to deepen and widen the channels of the Alazán, Apache, and Martínez Creeks, particularly at their confluences with one another and where they joined the river farther downstream. The heavily built-up San Pedro Creek offered a different set of challenges, leading Metcalf and Eddy to advise the city to acquire "a definite channel for that stream, and the removal of buildings on the stream banks that might interfere with a heavy flow of water."[25]

That the 1920 report paid substantially more attention to the river and the business district it flowed through than to the West Side creeks and adjacent neighborhoods was not by happenstance. The report's authors used two metrics, "hydraulic efficiency" and economic value, to guide their determination of what projects should be undertaken, and with what priority. By those standards, work on the creeks could be postponed: "*Improvements not urgent.* Owing to the comparative undeveloped nature of the territory through which these streams pass, there does not appear to be necessity for any large amount of improvement at the present time, although until considerable improvements are made the waters will spread over wide areas whenever flood flows occur."[36]

The report offered West Side residents this catch-22: properties in the district were too low in value to protect, but their value could not increase until flooding was prevented. Because there was no overriding reason to immediately improve the streams, flooding would continue unabated. These rationales gave city planners and politicians the green light to ignore this sector of the community until they perceived a compelling reason not to. As an example: while the report enumerates the critical interventions that should occur along San Pedro Creek, these were not as essential as those needed to straighten the San Antonio River. "The time when this work [on San Pedro Creek] should be undertaken will depend upon the growth of the city in population and wealth, and also upon the relative development of various parts of the city and the corresponding needs of local improvements." If other sectors were to grow faster in size and income, they should be protected before San Pedro Creek was improved. The same calculation held true for the Alazán, Apache, and Martínez Creeks. Their "betterment," and the "order of construction to be adopted on these creeks, cannot be determined at the present time," the report argued, because their "improvement is much less urgent." This diminished urgency itself was tied to a presumed fact: "The hazard of damage upon them is much less than in the congested [high] value commercial and industrial sections of the city." For Metcalf and Eddy, corporate and personal wealth determined public investment, a skewing of flood control policy reflected in their assertion that only when "the property along and adjacent to these streams becomes more valuable" would the channels be "modified as the conditions may then dictate to be most advantageous." This flood-protected future, according to the report, might not emerge until the 1940s, a prediction that turned out to be optimistic by three decades.[37]

As for the report's political reception, Metcalf and Eddy were well prepared. They knew that despite their careful analysis, and notwithstanding the pages of evidence-based graphs, charts, maps, and other illustrative material, the costs of sustained flood control would rattle local politicians. They did. Public officials did not doubt the $4 million price tag, but they were wary of bringing that figure to voters in the form of a bond. They did not dismiss the urgency, either, because Metcalf and Eddy did not equivocate in the report's executive summary. "We doubt if the citizens realize the ruinous loss which would result today with the present conditions of the river channels, from such a flood as that of a century ago (1819)," the authors wrote, a statement that was followed more ominously by a warning that the timing of any recurrence of such a mega-flood could not be predicted. "But that it will recur is certain. Therefore, with the rapid growth in value of property in the city, particularly in the congested value and commercial districts, it is imperative that this danger be recognized and that the work necessary to prevent serious injury from flooding be undertaken as rapidly as the financial resources of the city shall permit, lest when the flood comes it shall find the city unprepared and do ruinous damage."[38]

This portentous language and tough talk had an impact, if only in galvanizing the city council to pursue the report's low-hanging fruit. Chief among these were negotiations with the Guenther Mill to demolish its dam above the city and with other property owners about moving their buildings back from the banks. As for the proposed cutoffs, in 1919, prior to receiving the report, the city purchased property at the troublesome bend at Romana and Navarro Streets. In March 1920 it put this project out for bid, but the results proved more costly than the city had budgeted for, and it delayed the project. The 1921 flood brought additional delays to the Romana project, and the work of straightening the dangerous bend would not begin until 1925. Other projects outlined in the report undertaken in advance of the 1921 flood included construction of walls along portions of the river's banks, developing plans for dredging portions of the riverbed, and building more structurally sound bridges. Even these essentially minor alterations seemed to mark a major passing, the *Express* reflected: "The gradual changing of the San Antonio River from the stream of days gone by that went on regular tears at times of rain storms, recalls pointedly to the minds of the older residents the earlier days of the city during which fish abounded in the stream and two or three professional fishermen could be boasted of."[39]

The 1921 flood blew that lamentation out of the water. In the tragedy's wake there was no nostalgic harking back to some golden age when the river appeared to be tamer, with the people enjoying a more beneficent relationship with it, fishing rod in hand. Floodwaters were still sluicing across the streets when the *Express*, the *Light*, and *La Prensa* published editorials, columns, and articles favoring an upstream dam. The *Express*, which had lost electricity the night of the flood and could not put out an issue until September 11, led off with an imperative: "Build against another such disaster. Build a structure of flood prevention." Then it reminded its readers that eight years earlier it had championed a dam across the Olmos valley: "The cause and course of the smashing torrent of Friday night and Saturday morning demand this remedy be urged again on taxpayers and the city government."[40] The *Light*, knowing that factionalism might emerge to scuttle action on flood control, pressed for a united front: "Never before has there been such an urgent necessity for complete harmony among all the elements that enter into the municipal makeup of San Antonio. It would be another tragedy, and one superlatively reprehensible because it would be avoidable, if political considerations should in the slightest degree be given precedence over the common cause of quick recovery and future security."[41]

As for *La Prensa*, it trumpeted the pressing need for "Una cortina de concreto" to defend the downtown core. It based part of its argument, as did its English-language counterparts, on the city's long history of floods. Failing to correct the situation would only intensify the level of future damage. The newspaper predicated that claim on another: that San Antonio should follow the lead of such proactive cities as Chicago, Galveston, and Kansas City. After its incineration in 1871, over the next several decades Chicago "like a phoenix, from the fable...was reborn from the ashes more powerful than years prior." In the brutal aftermath of the 1900 hurricane that destroyed Galveston and killed six thousand, the Gulf Coast community built a seawall to protect itself from future storms. In 1903 much of Kansas City's river bottoms, home to its working-class poor and much of its industrial and commercial base, went underwater thanks to a rain-swollen Missouri River. Like Galveston, it constructed stronger levees, repaved its streets, and emerged as a more thoroughly modern city. The Olmos Dam, *La Prensa* predicted, would similarly transform San Antonio "into a shopping and industrial center of much greater importance than now."[42]

The chamber of commerce, public officials, and civic elite rushed to proclaim their fervent support for putting the river in a concrete straitjacket. Within days, the local Lions Club held a meeting dedicated to the question "What Will San Antonio Do?" For the fraternal organization the preordained answer was to immediately advocate for a dam.[43] Business owners and power brokers also met to discuss the need for flood control—what it would look like and how to pay for it. But most of all, each of these groups and their expansive networks were worried about how to keep the flood uppermost in the community's mind lest they forget, once the urgency faded, that more than seventy-five people had died, the West Side had been torn apart, the central business district had foundered, and that such death and disarray would happen again.

Mayor Tobin gave voice to these anxieties several days before the December 1923 bond election, which totaled a whopping $4.35 million, of which $2.8 million was dedicated to construction of the Olmos Dam and associated flood infrastructure. He had chosen the difficult route of a bond election because the city did not have access to federal funding that would later become available via the Flood Control Act of 1928. This legislation, in response to the devastating 1927 flood that damaged large portions of the Mississippi River watershed, underwrote the construction and rebuilding of levees along the Mississippi and its many tributaries. Los Angeles's controversial county-city bond elections in the late 1910s and 1920s are more analogous to San Antonio's situation. In neither region were bond elections a sure thing; indeed, the Los Angeles measures repeatedly failed. The possibility of a similar outcome made for a nervous Mayor Tobin.[44] "This election for flood prevention is the turning point in San Antonio's history. I hope everyone turns out and votes for a greater San Antonio. If we don't vote the bonds, we don't go ahead."[45]

The mayor was not the only one concerned about what a defeat at the polls would signify. In the week before the December 4 election, the *Express* ran reams of endorsements—pleas, in truth—from social and business clubs as well as professional associations, and these proponents were joined by bankers, furniture store owners, hoteliers, and insurance brokers. "Progress of San Antonio for years to come will be stopped if the flood measure fails," asserted Frederick Reutzel. Because she looked upon the flood bond "as a necessity for the continued welfare of the city," Mrs. J. Edward Dwyer argued that a pro-bond vote was a reciprocal act, part of a binding social contract: "It will be good for other residents of San Antonio to know that their homes and property are protected from water, and what is good for them is good for me." An apprehensive Charles A. Herff likened the central role of downtown commerce in nourishing the economy to that of the stomach. "Neglect your stomach and you will find that all other organs will cease to function properly. Neglect your business district and you will soon find, much to your dismay, that all outlying districts will cease to function and eventually bring on a collapse of our Dear Old San Antonio." Fretting supporters crafted countless letters to the editors; numerous advertisements promoted the flood control cause; the local newspapers wrote boosterish editorials.[46]

These pleadings fell on many deaf ears. Opponents were busy trying to undermine the momentum that had been building for the flood bond, and the brawl became so heated that Tobin linked the anti-bond forces to the Ku Klux Klan.[47] Election day, in short, was tense. The results were not the predicted landslide. While the other propositions that the bond would underwrite—funding for roads, bridges, sanitary and storm sewers, and other services—racked up impressive margins of victory, the flood infrastructure bond passed by a relatively narrow margin. Tobin appeared stunned, telling reporters he "felt a 'little blue' that the victory was not bigger for the bond issue." Yet he could not resist challenging the naysayers: "I am sure that when this great work is finished, the public will be sorry that all voters were not for it all along." His was a bit of wishful thinking: in a determined and well-funded bid to overturn the election, anti-bond forces brought a series of legal challenges, which the courts ultimately denied. By August 1924 San Antonio was cleared to release the bond funds and underwrite construction of the Olmos Dam; straighten, widen, and deepen the San Antonio River; and begin work on the West Side creeks.[48]

The dam and the river would absorb the lion's share of the bond money dedicated to flood prevention, more than $2.2 million of the $2.8 million voters had made available. That left $600,000 for the creeks, a substantial amount, given that they had never commanded any attention before. Mayor Tobin, who in 1923 had campaigned in part on the dire need to restructure these streams, wanted to make good on his promise, even to the point of increasing the funding for creek improvements. On August 22, 1924, he told the flood prevention committee charged with overseeing the city's flood projects that "he wanted it understood that it was proposed to spend $800,000 of the $2,800,000 voted for flood prevention on the Alazan and San Pedro Creeks." He added a critical qualification: "if it was determined that sum would be necessary."[49] Two days later, he walked back that larger commitment, reducing it to the original sum of $600,000; he conceded to the flood prevention committee the determination of whether that amount or a smaller one should be encumbered and acknowledged that 75 percent of these monies would be spent for labor.[50] Much of that went for salaries of those in the rapidly expanding flood prevention operations in the city engineer's office (at its height there were thirty employees). This was a necessary cost, to be sure, but that left an unencumbered $150,000 for the physical reconstruction of two creeks, whose objectives were significant: "All obstructions will be removed and the city plans to purchase the land immediately abutting on them. A flood zone will be created and no one will be allowed to build within this zone." Had this happened, the West Side would have been a much safer environment.[51]

While it is difficult to determine how much money was actually spent on West Side projects and what percentage of the work was completed, City Commission meeting minutes from 1924 to 1930, after which point flood control initiatives ground to a halt, suggest that much less was accomplished than was advertised. During that six-year period, the com-

Business and commercial buildings encroached on San Pedro Creek in September 1921. Note the high-water mark.

mission apparently released roughly $60,000 to purchase properties bordering San Pedro and Alazán Creeks. The former required most of the funds due to the extensive work of widening and deepening its confluence with the San Antonio River and to the higher property values upstream in the more densely developed urban environment San Pedro Creek ran through. Along the built-up sections, the city paid another premium: a set of flanking concrete walls were constructed to armor the creek as it coursed through a narrow corridor of buildings. Remaining funds seem to have been targeted on the confluence of San Pedro and Alazán Creeks, as the 1920 report had suggested. There were finally several big-ticket items charged to the "Alazan" account, but they would also have been critical to excavation work on San Pedro Creek and the San Antonio

River. The city bought two steam shovels (one receipt totaled $12,500), a Ford touring car, and a heavy-duty dump truck (prices were not recorded in the minutes), and for $187.50 the engineer's office snagged a "tree and log saw" that presumably received extensive use along the brush-choked West Side creeks.[52]

While recognizing that there may be a more complete ledger of costs and actions than those recorded in the meeting minutes, there is other evidence that indicates the West Side creeks were not substantially altered in the twenty-five years following the 1921 flood. During the night of September 26, 1946, and continuing into the early morning hours, a violent storm blew up over the watersheds of the creeks and river, dumping more than seven inches in less than eight hours. The Alazán, Apache, and Martínez Creeks carried almost as much water as they had in 1921; between 1 and 2 a.m., they ran over their banks, driving hundreds of people from their homes and killing nine, with half a dozen or more missing. The Ojeda family was particularly devastated. They lived just southeast of the confluence of Alazán and Martínez Creeks on Morales Street, and when the stormwater hit their home José Ojeda Sr. picked up his three-year-old, Rudolfo, and tried to carry him to safety. The turbulence knocked the father off his feet, and the boy was swept away. Six days later Rudolfo's body was found three-quarters of a mile to the south, lodged in another flood-ravaged house. His older brother, José Jr. (six), and stepsister, Olga Quinones (eighteen), trapped in their home, also drowned.

A USGS analysis noted that the flooding was a result of the creeks' structure: these "streams have small open channels with comparatively wide flood plains greatly encroached upon by residences, businesses, and numerous bridges." To finally remedy the creeks' flood-prone nature, the *San Antonio Express* editorialized, the city must "widen San Pedro and other creeks, construct dams and levees. One night's flood damage, as yet unestimated, would have gone far toward paying for the needed protection."[53] Those admonitory words would have been true anytime in the preceding quarter-century, and they further suggest that the Tobin administration had failed to deliver on its promise to alleviate flood conditions along the West Side creeks.

The Olmos Dam was a beautiful structure; its arches would be destroyed fifty years later in order to strengthen it.

At the same time, during the 1946 flood, Olmos Dam held back the maximum volume of water it was designed to impound. "The community saw its flood control works, installed after the 1921 disaster, operate helpfully," the *Express* commented. "Lacking the Olmos Dam, the downtown river cut-off, and widened creek-channels, the city might have suffered far-heavier damage."[54] Although the newspaper did not credit Mayor Tobin for his leadership in pursuing these flood prevention projects, he had a good understanding of their enduring value and of his legacy. Indeed, it seems reasonable to suppose that when Tobin stepped forward to speak at the dam's dedication in December 1926, he must have felt profound relief and satisfaction, having achieved what no other political leader in the previous century had been able to accomplish—bottle up the Olmos and shield downtown.

-{ FIVE }-

CONSTRUCTION PROJECTS

A potent symbol of change and renewal, the Olmos Dam was also a pro-jection of power—technological, economic, and political. As a "gravity" dam, the sheer weight of its construction materials resisted the horizontal pressure of water pushing against it. Its strength and resistance had been reinforced by an estimated 90,000 cubic yards of concrete further but-tressed by 418,000 pounds of steel. The dam looked massive, immovable.[1]

Its physicality helped some San Antonians begin to reorient to and reimagine the contours of their hometown. The dam would unleash a boom in suburban development on what would become known as the city's North Side, an ever-expanding periphery that has continued to de-fine San Antonio's residential pattern of growth for the upper and middle classes. This outward thrust came paired with a dramatic restructuring of the urban environment, notably the streetscape and skyline of the central business district. Below street level, and in the shadow of the new tall build-ings, the river too was transformed. Adding to work that had occurred seven years earlier as part of the River Park project, the once narrow and twisting river was further straightened, widened, and concretized. Once completed, this newly engineered terrain led to the development of the now legendary River Walk, one of the nation's great pedestrian byways. "Sanguine, indeed, is the future of the Alamo City," the *San Antonio Light* proclaimed in August 1924. "The recuperative energy she displayed after the disastrous flood in 1921 demonstrated beyond doubt that her vitality was not undermined." Given its enthusiasm, the newspaper would second the suggestion that the Olmos Dam was the single most important public works project in twentieth-century San Antonio.[2]

The Olmos Dam's remarkably generative impact did not mean that the dam's sanguinary benefits were universally accessible or evenly distributed. On the contrary, the dam's construction locked into place, with every barrel of cement poured into the wooden forms that gave it shape, many of the spatial injustices, social disparities, and economic inequities that would haunt the Alamo City's growth and development into the twenty-first century. These communal fissures in a sense were replicated in the Olmos Dam itself; this bulwark was not all it was cracked up to be.

Not that the celebrants at the dam's December 1926 dedication knew the structure on which they stood was compromised. Yet with a little imagination they might have glimpsed one of the environmental challenges that threatened the dam's security. As the happy throng returned to their cars, some probably took advantage of the structure's ninety-foot elevation, gazing north and west across the construction-scarred floodplain to the low hills framing the horizon. What they were looking at was a portion of Olmos Creek's thirty-two-square-mile watershed, a rolling terrain that the Payaya and their ancestors had utilized for millennia (some of the physical evidence of their lifeways had even been unearthed during the dam's construction).[3] Now, however, that stream-fed landscape had been converted, quite literally, into the stomping grounds for a number of dairy and cattle operations that fed local consumers. The location of these ranches had been critical to one of Metcalf and Eddy's recommendations: the city should contract with ranchers who lived along the upper reaches of Olmos Creek, and who crucially owned telephones, to erect and manage US weather gauges on their properties; during major storms they were to keep a close eye on their gauges and to alert public officials to prepare for floods when major storms pounded the region.

Albert Maverick was among those recruited. He had calculated, after emptying his gauge three times during the September 1921 storm, that more than 14 inches of rain had fallen on his property, which lay within the upper reaches of the Alazán Creek watershed. Others located on the farthest northwestern edge of the Olmos Basin recorded between 15 and

17.5 inches. The sheer amount of rain, when juxtaposed to the 7 inches that fell over the city proper, helped explain why the flood had been so severe.[4]

Its severity and suddenness were related in part to the livestock these storm watchers owned, and which since the mid-nineteenth century had heavily grazed the hills that ringed the watershed's slopes, ravines, and creek banks. Over the succeeding decades, the number and impact of these animals swelled as San Antonio's appetite for dairy products and meat grew with the rapid increase in population. Two references to the change in the watershed's ecological integrity give a sense of the transformation that occurred between 1869 and 1921. In July 1869, a thunderous storm exploded over the Olmos Creek / San Antonio River basin, dropping 14.5 inches of rain. Seeking an explanation for why the downstream damage was not as intense as it might have been, the *Express* argued that one merely needed to look upstream. "The reason we were spared a disastrous overflow from the recent unprecedented rain storm is due to the thick mat of vegetation that covers our valley, thereby impeding the sudden rush of water from the hills, retaining a great deal and graduating the flow of the surplus."[5]

Fifty years later that protective groundcover had been trampled, a process that World War I had intensified. With pricing incentives from the federal government and a resultant spike in demand for meat, hides, and wool, ranchers all along the Balcones Escarpment, which included the San Antonio hills, and higher up on the Edwards Plateau, increased the number of cattle, sheep, and goat in their herds. Dairy farmers were also induced to build their stocks. As a result, animal products surged into the marketplace. San Antonio's many military bases and training facilities were important consumers of such productivity, and the city as well was an important railroad hub, which dispatched agricultural goods and services across the nation. The local grasslands and waterways paid a heavy price for this agricultural boom. Grasses, forbs, and brush were mowed down; stream banks collapsed; and erosion led to silting, which led to reduced stream flow and poor water quality. These environmental damages had important downstream consequences.[6]

One impact was manifest in the 1921 flood's swift downhill run, in San Antonio and the other river-hugging cities to its north. That was the crux of the argument of W. W. Ashe, a US Forest Service scientist, senior forest inspector, and secretary of the National Forest Reservation Commission. Since the 1890s, Ashe had been analyzing the influence of deforestation and grassland despoliation on the increase in the number and severity of floods. In late August 1921, as part of his fieldwork for the National Forest Reservation Commission, which was established in 1911 to purchase upper watersheds to conserve and protect these valuable lands and their water resources, Ashe traveled to Texas. His goal was to assess the condition of the upper reaches of the Colorado, Guadalupe, Comal, and San Antonio Rivers and to determine whether portions of this rugged terrain, and of the Edwards Plateau, might qualify for inclusion in the national forest system. Ashe's expertise in hydrology and watershed dynamics would receive a shocking, real-world test. He arrived in San Antonio on September 9 just as the heavy rains began to fall and was trapped in the subsequently flood-ravaged city for several days before traveling to Austin to speak to the State Board of Water Engineers. It was in the capital that an *Express* reporter caught up with him and asked for a big-picture perspective on why the Alamo City and its wider region were so vulnerable to floods.[7]

With an attention-grabbing, front-page headline—"Cities Will Continue to Be Devastated Unless State Acts"—the *Express* published its interview with Ashe on September 16, even as San Antonio was still reeling from the disaster. In the piece, Ashe drew parallels between the physical geography of the Edwards Plateau and his home ground in the southeastern states. The "streams of Texas are erratic and exhibit the same character of flow as those at the southern end of the Appalachian Mountains," he explained, which was a consequence of "the enormously heavy rainfall at irregular intervals and rapid run-off on account of steep slopes."[8] What complicated these natural processes, and thus was a driver of the devastating floods that routinely wracked each region, was the human imprint. In western North Carolina, it was logging and grazing. In central Texas, so-called "cedar choppers" clear-cut large swaths of juniper stands for fuel, fencing, and railroad ties, leading William Bray, a federal forester like Ashe, to predict that little could be done to stop the intense logging: "So long as the small owners depend in large measure for their income upon the sale of wood, the temptation will be strong to denude rough, thin-soiled hillsides which would far better be kept with a protective timber covering," Bray had written in 1904.[9] Compounding this serious environmental problem, Ashe declared in 1921, was a "greed for land" that led those ranchers whose property abutted the escarpment rivers to construct levees so close to streambeds that they robbed the catchment basins of their ability to act as sponges, thereby intensifying downstream damage. More impactful still was that unmanaged and excessive grazing in upland pastures and meadows had stripped the already thin soil of vegetation so that even modest storms could generate major floods. It was no wonder that the deluge on September 9 and 10 had produced such dire consequences in San Antonio, New Braunfels, San Marcos, and Austin.[10]

-[*129*]-

The resolution, Ashe advised, was to manage at a landscape scale. Engaging with the watershed as a watershed, he assured readers of the *Express*, would provide quantifiable long-term benefits. The Forest Service had demonstrated how effective this intervention could be in its watershed management across the country. Just as the federal agency was doing in North Carolina, the Mississippi River Valley, and Colorado, the key to controlling floods in Texas was the "protection of the forest cover in the central portion of the State in gorges, along the flood plains, on mountain slopes and in ravines." The "fissured limestones of the Edwards Plateau" posed significant dilemmas that careful stewardship could resolve. "The woodland and herbaceous cover of this region is too light to protect the surface from erosion," Ashe noted later, "but it is possible, by adequate regulation of grazing, and by better protection of stream banks, if not to reduce erosion at least to prevent its further increase."[11]

Failing to institute rigorous regulations would have ongoing and disastrous consequences. Ashe pointed to the heavy erosion along the Colorado River, for example, that had led to rapid siltation. In the past, this siltation had compromised the stability of the Austin dam. It gave way during major floods in 1900 and 1915 that generated a cement-like slurry of trees, rocks, gravel, and sand that destroyed the structure before devastating the Texas capital. To escape that fate and to maximize the effectiveness of dams along the Balcones Escarpment in central Texas—including San Antonio—Ashe urged the adoption of a hybrid approach to flood control that combined careful, deliberate management of the headwaters' ecosystems and, where appropriate, the construction of downstream infrastructure such as the Olmos Dam. Ashe's call for hybridity was embodied in the different professions—foresters, rangeland specialists, hydrologists, and engineers—that he argued must cooperate to build more resilient and secure river systems. Should this cooperative planning and implementation occur, Ashe assured still-dazed San Antonians, "storm waters, in place of being an agency of calamity and destruction...will become one of the most potent and permanent of the resources of the state."[12]

San Antonio's leaders did not heed Ashe's advice, if they even took notice of it. Instead of developing a master plan to mitigate flooding within the larger Olmos watershed, local politicians and the civic elite followed the lead of Metcalf and Eddy (and prior generations of flood control advocates) by focusing squarely and solely on what they believed was the keystone structure, a dam that would seal off that basin's dangerous waters. Yet had they followed Ashe's suggestions, upstream land management might have better protected the dam after it was completed. It, after all, needed that protection, too. That is what a pair of scathing, post-construction engineering reports on the Olmos Dam revealed. Each was submitted to the dam's contractor and to public officials in spring 1927, and each identified several serious structural flaws. Had the reports' recipients paid attention to these criticisms, they might have been less confident when the dam's concrete had set, and the public dedication been concluded, that their work was done.[13]

The reports were the result of a stipulation in the city's contract with Col. S. F. Crecelius, the dam's project designer and engineer, and McKenzie Construction, the builder. The contract required that upon the dam's completion, two outside engineering teams would evaluate whether the dam had been built "in accordance with its plans and specifications."[14]

One of the reports, written by a California consulting engineer and of which there is only an unsigned copy extant, was sent directly to A. J. McKenzie, vice president of the eponymous contracting company. Because he was a long-distance consultant, this engineer relied on a close examination of the contractor's reports as well as an array of photographs that McKenzie Construction had taken at each significant stage of construction. Although the engineer acknowledged that it would be "assuming too much and rather unwise for the writer to make a definitive statement as to the suitability of this dam as designed without seeing the site and investigating very carefully the foundation conditions," he was unsparing in his analysis.[15]

-{ 131 }-

The first problem was the foundation. "If I understand your photographs correctly the bedrock was cleansed of overburden, but no cut was made in the rock itself," the engineer stated. Should that be an accurate reading of the situation, then the structure was thereby weakened: "We have never built a dam without cutting into the bedrock over the whole bearing area to a greater or lesser depth depending on the suitability of the rock for bearing the load and for water tightness, and also keying the dam into the foundations." Equally concerning, and a further clue that the bedrock had not been cut, was that the dam as constructed did not follow basic engineering guidelines. "On all our dams the foundations have been prepared so that the pitch of the bedrock was toward the reservoir…and was terraced or stepped to give further resistance against sliding and also to make it more difficult for water to follow the line between bedrock and concrete." Because Crecelius's design had left the "general pitch" tilting downstream instead of up, the dam was compromised.[16]

Other flaws included an incomplete drainage system, which implied that the designers had not recognized that "it is very vital to prevent there being any accumulated head under the structure"; "head" in this context refers to pressure on the structure. This point was underscored in the report's note that the site also lacked a cutoff trench on the upstream face to "intercept seepage." More troubling was the lack of a spillway, "which seems to us incredible, particularly because the only outlet is so restricted and so low compared with the surrounding reservoir floor that it would easily and quickly be clogged if there was any rush of water." Maybe the lack of a spillway would suffice if the Olmos Creek could be counted on to fill up the reservoir "in a slow, quiet way," but in California a dam with only a single outlet at ground level "would probably be clogged in the first two hours of a storm." Should a debris-choked Olmos Creek roar down its watershed as it had on countless occasions—especially in 1921, a situation that Ashe warned had twice toppled the Austin dam—then "it would be only a matter of time until the dam must overflow." That led to this chillingly dispassionate conclusion: "I should judge this was not intended and might make serious trouble."[17]

Blunter still was a three-page, single-spaced, legal-sized report by a trio of San Antonio engineers that was submitted directly to Mayor Tobin and the city commissioners. C. H. Kearney (who once had been the city engineer), A. Y. Walton (like Crecelius, a retired officer in the US Army Corps of Engineers), and R. B. Huffman (a consulting engineer) conducted an on-the-ground investigation of the dam site. Their walking inspection, when combined with their careful evaluation of construction records, including blueprints, photographs, and other material pertaining to the dam's day-to-day construction process, was as thorough as their conclusion was unsettling. "In making this inspection, we observed certain conditions in connection with this project which we, as professional engineers, sincerely believe render the structure impractical and unsafe." First among these was the foundation's many deficiencies. They too scored the designer's failure to correctly build the foundation. Because they were local, they knew that where the dam was sited was rife with springs that poured out of the limestone bedrock; indeed the project had been delayed because workers had to spend days trying to pump water out of some of these underground fissures before they poured any concrete.[18] Given that "a greater portion of the length of the dam...was located on a honeycomb rock, filled with holes and seams of unknown depth," and that Crecelius had failed to include appropriate anchorage, barriers to seepage, cutoff walls, a drainage system, and the correct upstream tilt to the structure, the local assessors' judgment was unequivocal: "No part of this, or any other dam, should rest on the surface of a foundation such as this."[19]

Missing design features stipulated in the contract—among them aux-
iliary gates in addition to the sluiceway—compounded the errors these
engineers tabulated as they did their site inspection. They paid attention
to the ten-foot-deep channel leading to the sluiceway gates that normally 〔 133 〕
would remain open to allow water to flow freely downstream. Because the
gates' bottom was level with the channel's bottom, "we believe that the
first flood and all subsequent floods that come along will bring down so
much drift and silt that the racks will soon be choked to such an extent
that only a small fraction of the water intended to go through the conduits will be able to pass the racks and the whole object of the reservoir
destroyed." For evidence of the high likelihood of this happening, the engineers pointed to the "great amount of drift that lodged against the International–Great Northern trestle crossing the Olmos a short distance
above the dam during one of the recent floods. These 14-foot openings
were seriously blocked by all sorts of drift, including large trees, little of
which could possibly be passed through the 4.5-foot openings in the racks
in front of the [dam's] outlet tunnels."[20]

The lack of a spillway also baffled the San Antonio engineers as it had
their California colleague. This absence alone was such a "very serious
mistake" that it endangered the "safety of the structure and of the inhabitants of the community." But perhaps their most disturbing finding
concerned the quality of the concrete that constituted the gravity dam's
weight. After going through the daily reports at the mixer, the San Antonio consultants concluded that the contractors had cut costs by changing
the ratio of cement to limestone and crushed stone to produce "a very
lean concrete even when the materials used as aggregate are of the very
best and are properly graded." Unfortunately these materials had not been
routinely and uniformly graded: other findings revealed that "the limestone screenings in many instances were of poor quality and that due to
the fact that the aggregates furnished by the city were of such irregular
sizes that it was necessary to change the proportion of the concrete very
often, said changes being made as often as three times an hour and ten
times in one day." The result was that the concrete was uneven in quality,
not as durable as it should have been, and less resistant to water absorp-

tion.[21] However colossal the Olmos Dam looked, its bones were weak. Decades later, the city would be fortunate that the September 1946 flood did not overtop the structure, though it came within a foot or so. Had the floodwaters gone over the dam, there is a strong possibility that the resulting churning waters would have undercut the structure. Low-lying residential neighborhoods to the dam's south and San Antonio's downtown core would have been destroyed.[22]

Even if that worst-case scenario occurred, some sectors of the city would have been spared, including the classic late-nineteenth-century streetcar suburbs on higher ground a mile or two north of San Antonio—Laurel Heights, Tobin Hill, Monte Vista, and Alamo Heights. Their names conveyed their physical elevation, a coveted form of flood insurance. Because these neighborhoods were so desirable, they contained the city's priciest real estate. Ruth Dubinsky Kallison, like her husband, Morris, had grown up in Laurel Heights; the young couple initially lived in an apartment in the same neighborhood. So removed were they and their neighbors from downtown that Ruth Kallison's first realization that San Antonio had been battered as a result of the September 1921 flood came when an out-of-town friend called to make sure she was safe. "'Of course we are all right, why not?' And they said, 'there's a terrible flood in San Antonio and we just heard about it.'" Kallison drove downtown with some of her friends and watched "the water rush into the buildings." "A lot of the buildings were ruined," she said. "Many people lost everything. . . . San Antonio was not prepared. We did not have the protection we needed for a flood."[23]

In search of added defense against losses caused by flooding, many well-off San Antonians were attracted to a cluster of automobile suburbs platted simultaneously with the announcement that the city intended to build the Olmos Dam. Suddenly the low, oak-studded hills just outside San Antonio's northern limits looked mighty attractive. The *San Antonio Light*, in a series of prescient editorials in August 1924, predicted a new land rush. "Take a compass and draw a circle on the map with San Antonio as the center and a radius of ten or fifteen miles," it urged, in "Buying San Antonio's Hills." Once drawn, it encouraged readers to "get into your

automobile and drive out every improved highway in every direction at
your leisure. Count the number of hilltops you see in the area thus covered." After the circuit had been completed, "you will have a list of one of
the great natural resources of this city," once ignored but now considered
priceless. "Most people are unaware that a hilltop, formerly considered
waste land, today has a value out of proportion to the surrounding lands."
Part of the draw was that there were few of them, thus increasing their
price. Another part of their appeal was their perceived status: "The country homes of the future are to be hilltop homes." And another, if unspo-
ken, benefit of staking a claim on the hilltops in this flood-prone region
was that high ground equaled safety.[24]

Who could afford this elevated terrain was suggested in the title of
another *Light* editorial: "We Are Going to the Suburbs." The first-per-
son-plural encompassed a narrow range of the city's population: "those
favored workers and professional folks who can regulate their hours going
and coming," who have the disposable income to daily purchase tickets on
a bus or to cover the cost and maintenance of a car. The latter was a new
form of mobility "that is going to exert a hitherto unknown influence on
man's place of residence." For this privileged population, the suburban
future was ever expanding. "At the expense of being called dreamers, we
may picture the suburbs of San Antonio extending away many miles into
the blue haze of the hills and lands, that are today, rough farm and ranch
lands" and that in due course will become "country homesites."[25]

One real estate developer who turned those dreams into reality was
H. C. Thorman. Shortly after the dam's construction was announced,
Thorman purchased sixteen hundred hilly acres that abutted the future
project on its west. It would prove to be his largest and most exclusive
development. Previously he had built middle-class neighborhoods in the
Highlands and Fredericksburg Road areas, southeast and northwest of
the downtown core, and Country Club Estates, middle-class housing that
bordered the San Antonio Country Club and Fort Sam Houston to the
city's northeast. In 1925, after less than ten years in the business, Thorman claimed to have built more than twelve hundred homes, a substantial
number at a time when most suburban development in the United States

The Park Hills Estates neighborhood was built for the elite.

was on a much smaller scale. Further evidence of Thorman's successful rise was the home he constructed for his family in the posh suburb he was creating that in time would be called Olmos Park. Set amid an oak grove, complete with a two-and-a-half-story glass dome foyer, the colonial-style mansion was reported to have cost in excess of $50,000.[26]

Thorman's success was keyed to the car. He anticipated that the Olmos Dam and the road that ran across it, offering a "new crosstown thoroughfare linking Alamo Heights on the east with Laurel Heights on the west," would redirect traffic patterns and make his investment more accessible, desirable, and profitable.[27] No surprise, one of his first advertisements for Olmos Park features a surge of automobiles rushing west across the dam from Alamo Heights toward a banner emblazoned with the words PARK HILLS ESTATES, a subset of his subdivision. Thorman

| FOR SALE REAL ESTATE | FOR SALE REAL ESTATE | FOR SALE REAL ESTATE | FOR SALE REAL ESTATE | FOR SALE REAL ESTATE |

WHY LIVE IN THE MUD?

$1 DOWN $1 PER WEEK
NO MUD   NO INTEREST

MITCHELL PLACE

DEDICATED TO WORKING PEOPLE

LOTS $175 TO $300 EACH

Every street in MITCHELL PLACE has already been completed and when you see the fine gravel streets you will say "no more mud for me." Mitchell Place is the addition where you can live on your lots while paying for them and build any size house you wish, be it large or small.

WHY NOT BUILD NOW

AND SAY MAN! You don't know how easy it is to get a roof of your very own over your head, but if you will come to Mitchell Place we sure can help you. This sale has been opened only a few days and already lumber is being delivered on the lots for homes by men that "took time to investigate" our proposition. Will you do the same and—

MAKE YOUR RENT MONEY PAY FOR YOUR HOME

STREET CAR ONE BLOCK—GRAVELED STREETS—ELECTRIC LIGHTS, PHONES, ETC. TO THE PROPERTY.

WHITE PEOPLE ONLY.   WHITE PEOPLE ONLY

TAKE SOUTH FLORES CAR to Mitchell Street. Walk one block east to the property.

Strike while the iron is hot—come early—WE WILL BE THERE ALL DAY SUNDAY and will take pleasure in showing you the property without the least obligation on your part.

N. S. DICKINSON & CO.
"LOT SPECIALISTS"

Crockett 6437   218 Losoya Street

Mitchell Place was a working-class subdivision south of the downtown core.

stands under the banner, waving the cars on, ready to make a deal.[28] The automobile, which determined the community's spatial design, was indispensable because Olmos Park was served neither by a streetcar nor by anything so pedestrian as a sidewalk. That dependency could have led Thorman to replicate the gridiron pattern that characterized most of San Antonio's streets, a pattern that would obscure the natural contours of the land for the sake of convenience and efficiency. Instead he designed Olmos Park to rein in the car and to provide a sharp break—visual and physical—from the relentless urban geometry. After giving "careful thought to the planning of these Estates [such] that the great natural beauty and splendid position of this property would be enhanced," he embraced what he called the "parkway system of development." This system, which drew upon the landscape practices of Frederick Law Olmsted and J. C. Nichols, consisted of broad avenues and drives that wind among the native oaks and roll gently over the hills. This was a restful environment, a natural tonic for the harried urban dweller.[29]

The community was further insulated from the surrounding city by natural and manmade barriers. To the east and north of Olmos Park lay the Olmos Creek flood plain, land forever uninhabitable now that the dam had been built. To the south lay additional parkland and a quarry; portions of the western section bordered the International–Great Northern (later Missouri Pacific) railroad tracks. Each obstacle prevented the encroachment of undesirable development and, when combined with the centripetal force that the "parkway system" exerted on movement in the community's interior, set Olmos Park apart from its environs.[30]

Social exclusivity was intrinsic to this new development. As with other upper-class suburbs developed in the early twentieth century—such as Roland Park in Baltimore, Houston's River Oaks, and the Country Club district of Kansas City—Olmos Park adopted a restrictive covenant to ensure racial segregation, high property values, and the perpetuation of these features into the future. Thorman and other developers had turned to restrictive covenants when the US Supreme Court had declared segregation by zoning illegal in 1917. Nine years later the high court effectively sanctioned restrictive covenants when it refused to hear *Corrigan v. Buckley*, and consequently the popularity of these covenants soared. The racial restrictions in the Olmos Park covenant reflected a national trend, and a close examination suggests the complex role such documents could play in the creation of segregated, elite suburban bastions.[31]

Like elites throughout the United States, those in San Antonio sought refuge from the steady advance of the city and its presumed social ills and racial demographics. Thorman addressed these concerns through the restrictive covenant he put into effect for Olmos Park in January 1927, which both shaped the community's demographic character and determined land-use patterns and housing options. The 1921 flood had exposed the city's sharp racial, ethnic, and class divisions, and Olmos Park capitalized on, extended, and perpetuated their impact.

Racial codes and economic restrictions in turn played an important role in the selling of Olmos Park. They were prominently displayed in newspaper advertisements designed to lure those who valued such protections and the exclusivity they provided. The covenant made it clear that Olmos Park was for whites only: no portion of the property could be "sold, conveyed or leased to any person who is not of the Caucasian race." Any violation of this stipulation was punishable by law and would "work as a forfeiture of the title to the particular subdivision of property," a clear indication that the community's right to racial segregation superseded the rights of individual property owners. The covenant both applied to the original owner and was made "running with the land" and therefore applied to and was bound upon "the grantee, his heirs, devisees, executors, administrators, successors or assigns." The restrictions would "forever stand good," Thorman declared. The covenant, like those adopted across the nation, acted as a "social compact," in historian Thomas Philpott's words, that "symbolized and guaranteed community solidarity," a solidarity that erected and enforced racial boundaries.[32]

Other stipulations ensured the community would be only residential in nature. As the warranty deed for Olmos Park Estates puts it: "Neither the [original land] grantee nor any subsequent owner or occupant of said property shall use the same for other than residence purposes." Underlying this seemingly innocuous statement lay a host of concerns that a late 1926 advertisement for Park Hills Estates addressed: "The property has been blanketed with restrictions in order to assure development of the character that this realty warrants. It has been restricted against apartments, hotels, and business of any character." Another advertisement made it clear that the restrictions were "sensibly designed to protect your home and every home from the encroachments of inferiority."[33]

Thorman projected this sense of exclusivity and refinement in every newspaper advertisement. One ad features a white, conservatively dressed, mature audience in the foreground. One man wears a monocle, most other men are in evening dress or business suits, and the women are as fashionably attired as their husbands. In the background, facing the audience, stands Thorman, wearing an evening jacket with satin lapels. He gestures toward a map of the suburb's site plan set in a scrolled frame accented by velvet drapes. The vista reflects the richness of its environs. Amid rolling hills, substantial homes unobtrusively line the community's sweeping drives. Only people of substance would purchase property in this community, the advertisement asserts, and their homes' magnificence would reflect their heightened social status.[34]

This class-based appeal was directed toward San Antonio's commercial and business elite, who lived in areas that may have lacked effective covenants, were inundated by the 1921 flood, or were threatened by expansion of the central business district. Britain R. Webb, a vice president of the City Central Bank and Trust and regional manager for the Buick Motor Company, and one of Olmos Park's earliest residents, pulled up stakes from the once exclusive West French Place, which had sheltered a number of leading citizens in the early 1900s. His new neighbors included Joseph Edwards, an auditor for the city's Packard distributor; Clarence Gardner, treasurer of Mountjoy Parts Company specializing in automobile engines; J. Benjamin Robertson of the Luthy Battery Equipment Company; and Raymond Woodward, owner of Morgan-Woodward, a local Ford dealer. They gave explicit meaning to the concept of an automobile suburb.[35]

An added draw for these and other new residents was that Olmos Park was situated north of the city limits, enabling them to avoid city taxes and direct participation in San Antonio's public affairs. Yet perhaps most compelling was that the community rose above and behind the Olmos Dam. Even those acres bordering the Olmos Creek floodplain were elevated enough to avoid overflow. A marker of its safety is Contour Drive, which follows the stream as it curves along the neighborhood's northeastern flank and is said to demarcate the level of a hundred-year flood, a claim that the 1946 flood, the largest since 1921, tested. So recalled Frances Rosenthal Kallison. She and her husband, Perry Kallison, had moved into a limestone and red-tile-roof house fronting Contour Drive, and it was from this dry vantage point that they watched impounded floodwaters fill the Olmos Basin, spotted a neighbor rounding up his horses from a motorboat, and witnessed the flood's high-water mark—swirling water lapping on the street, a macadam beach. "My children had two ducks named Elmer and Sadie," Kallison remembered. "The children were cackling with glee and delight when they looked out the window and saw Elmer and Sadie having one grand time disporting themselves up and down Contour."[36]

On drier mornings, Contour and the other meandering roads in Olmos Park were filled with cars whose drivers were heading downtown. Among them was H. C. Thorman, who motored from his colonnaded southern manse while brothers Walter and Richard Negley pulled away from their commodious Spanish colonials, as did Carl Newton. Those inhabiting slightly more modest limestone-clad homes made the same commute. They departed from a neighborhood whose aesthetics were designed to evoke an Old World elegance and charm only to arrive downtown, where, more often than not, they were employed in one of the new modern office towers. Their daily movement through space; the motorized vehicles that allowed them to cover the four miles between suburban domicile and urban workplace; the newly paved, widened, straightened, and/or extended streets; and the high-rises where they spent their working day were not experiences unique to San Antonians.

Their elite contemporaries in New York and Chicago, Dallas and Houston, were similarly housed, mobile, and occupied. One sign of their shared modernity was their diurnal commute. Another, even more obvious because of its signature size, was the skyscraper in which increasing numbers of them could be found daily. "Surely more than any other type of building," Paul Goldberger observes, "the skyscraper is both quintessentially American and quintessentially of the twentieth century," a bold force "as powerful in its ability to transform the urban environment in its time as the automobile." Its power was a partial result of the tall buildings' offering a sharp break from the once-cluttered streetscape, "the inconveniences and congestion of the traditional city. Skyscrapers would provide grandeur and views and also clean streets and efficient communication." As enhanced mobility and the straight lines of this new architecture produced a more legible and geometric urban form, they altered how people perceived what made a city a city. No longer just "an agglomeration of small- to medium-sized buildings, made urban by their closeness," a city,

courtesy of the skyscraper, was redefined on the "basis of size," Goldberger argues. It "showed its might by how many buildings it had and how many people were in it.... Our very notion of what cities were was
changed forever."[37]

These transformations held true in San Antonio, which experienced a building boom like no other in its history. The impetus was clear in this *San Antonio Light* headline: "Flood Prevention Means $12,000,000 in Building."[38] From the moment the 1923 bond was announced, and with the Olmos Dam now a given, local and national investment spiked. This "allowed for the construction of monumental structures and buildings designed for advertising purposes." Their monumentality and branding were made visible in the Sunday real estate sections of the *Express* and the *Light*, which teemed with stories about new bridges being built, streets under construction, hotels and office buildings planned or underway—a bustle of activity that intrepid photographers snapped from atop each new edifice. This visual record offered San Antonians a new panorama and perspective, a bird's-eye view as the city zoomed into the sky.[39]

Land values soared even in long-neglected sectors. In February 1927, for example, the *Express* tracked a spurt of real estate sales along East Houston Street and several of its intersecting arterials. "Property values have more than trebled along some of these streets within the past four years." Land that in 1922 and 1923 had been offered at $750 per front foot "would now cost from $1800 to $2000 per front foot," a boost that also would be realized on North Alamo Street and Avenue E "when the projects that have been announced have been completed." The earlier construction of anchor projects drove the surge, projects including the magnificent Scottish Rite Temple (1924) and the strikingly V-shaped Medical Arts Building (1925). Future projects including a new federal building and post office (completed in 1937) and a new headquarters for the *San Antonio Express* (1929) also contributed. These projects, when combined with street improvements, caught the eye of investors and speculators looking for bargains as prices in the central core became astronomical.[40]

Shortly after construction began on the Olmos Dam, an estimated $10 million flowed into the downtown real estate market. That was the first of

A stereograph of the Medical Arts Building.

a succession of waves of capital that pooled in the wider business district. Including those along East Houston Street, between 1922 and 1930 more than a dozen major structures were built. They stood out for their size, materials, and design, features that required additional expense. Almost all are now on local, state, or national registries of historic places, serving as iconic structures that continue to define the city's streets and skyline. When completed in 1922, for example, Frost National Bank (now a municipal office tower) was the city's tallest building at twelve stories and dominated Main Plaza. Its Fort Worth–based architects, Marshall R. Sanguinet and Carl G. Staats, taking their cue from the owner's business, "incorporated round bas-relief panels depicting U.S. coinage set between double-height arched window openings."[41] Local architects Atlee B. Ayers and Robert M. Ayers were brought on to design the Plaza Hotel (1927; now Granada Homes) situated on what had been Bowens Island (no longer an island, courtesy of a river-straightening project). Like many of the homes in Olmos Park that Ayers and Ayers were simultaneously planning, the hotel reflects the firm's "mastery of Spanish Colonial Revival design."[42] Also located on the former Bowens Island was the Federal Reserve Bank (1928), a branch of the regional Federal Reserve Bank in Dallas. As befit its satellite status, the building was modest in size but made up for its lack of height with a stately granite exterior and polished marble

The Smith-Young Tower, 1929.

interior. The former bank's "quiet dignity is highly appropriate for the building's present tenant, the Consulate of Mexico."[43] That same year, the twenty-one-story Milam Building (1928) was completed; its height and bulk overwhelmed the eponymous plaza. Its early occupants, headquarters for many regional oil and gas companies, may have been lured by its humidity-busting amenity; the building was "the world's first high-rise structure originally designed with full mechanical air conditioning."[44]

In 1929 three more buildings pushed the city's skyline higher still. Boosting "what is perhaps the city's finest collection of terra-cotta ornamentation," the twenty-three-story Nix Professional Building, whose general manager was an early resident of Olmos Park, houses a hospital and medical offices and is set along the banks of the San Antonio River.[45] At twenty-four stories, the Alamo National Bank rose above Commerce Street; a Chicago firm employed a restrained classicism to underscore the financial institution's solidity.[46] The most extravagant and shadow-casting structure was the Smith-Young Tower (now the Tower Life Building), which at thirty-one stories rises 404 feet above Navarro Street. Its striking octagonal shape, dramatic integration of "brick and terra-cotta detailing with Gothic-spirited ornamentation replete with gargoyles," and floodlit upper floors were designed to make a bold statement.[47]

On a smaller scale, so did a series of cultural buildings, including Municipal Auditorium (1926), yet another Spanish Colonial Revival structure in a city filled with them; two exotically, riotously decorated movie palaces, the Aztec Theater (1926) and Majestic Theater (1929); and the San Antonio Public Library (1930; now the Briscoe Western Art Museum). Fronting the river's Great Bend, the museum is arguably the best local example of "modern classicism."[48]

All of these structures—public and private—remade downtown. As they lifted the economy, they created new zones of opportunity, interaction, and engagement. Yet this upgraded urban landscape also reinforced entrenched zones of exclusion. Event ticket pricing separated movie- and concertgoers by class. Jim Crow laws determined where they sat, and who with; these same laws also determined who could check out books from the downtown library. How and where work was done was reconfigured in the remaking of downtown, as well: tall buildings concentrated their occupants by occupation. Medical and healthcare practitioners, like those in banking and finance, natural resource development, and other white-collar specializations, found office space with their professional peers. Emblematic of this sorting impulse was the Builders Exchange Building (1925) on East Pecan Street. With considerable investment from the National Association of Builders Exchanges, the ten-story Gothic Revival structure leased offices to architects, contractors, surveyors, and suppliers, trades in high demand during the 1920s boom in residential and commercial development. Those who promoted these projects also

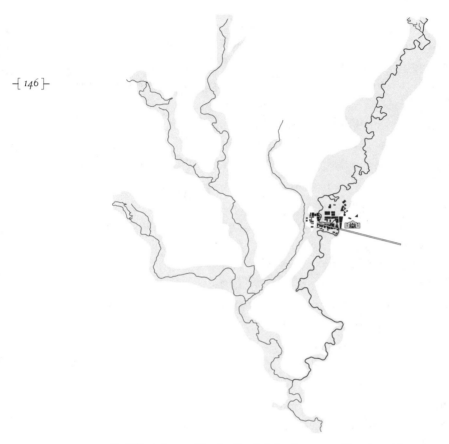

Buildings located in the floodplain, downtown San Antonio, 1921.

built a high-rise, their "own home." The ten-story Real Estate Building (1929) at the corner of Saint Mary's and Martin Streets sported a brick and limestone façade with terra-cotta and art deco detailing; it housed the local realty board, property managers, escrow companies, and real estate agents.[49]

The siting of these structures reflects another crucial component of the city's uptick in upscale construction. Each building was located within the expansive floodplain that the 1921 inundation had violently reclaimed. Water that raced out of the river's banks in Brackenridge Park and angled south down the aptly named River Avenue flooded the same section of

Apache Creek, showing the systemic disparities that defined San Antonio.

East Houston Street that by the mid-1920s would become such a hot secondary real estate market. Saint Mary's Street coursed with five to eight feet of water, Bowens Island was submerged, and the major east–west avenues—Houston, Commerce, and Market—became one churning sea spreading from Alamo Plaza on the east to Main Plaza on the west, a fetid swamp of dead animals, effluent, garbage, and other debris. The new skyscrapers stood like a profane gesture to Mother Nature. With concrete and steel, human beings had triumphed over her destructive powers.

Left untouched by the upscale building boom were the desperate conditions—poor housing, marginal employment, little to no access to potable water or sewer systems, minimal schooling—that plagued inhabitants of the West Side before the flood and would do so long after. These inequities, at once social, economic, and spatial, are exemplified in a 1939 photograph from the camera of Russell Lee, one of the Farm Security Administration photographers who roamed the country to document the impact of the Great Depression. In the foreground stands a dilapidated shack in a dusty clearing just west of San Fernando Cemetery where so many victims of the 1921 flood were buried. On the eastern horizon rises the Smith-Young Tower and Alamo National Bank, twin towers that epitomized the downtown core's explosive growth in the previous decade. The contrast could not be starker: the dam and downstream flood control, which offered no protection for those living on the crowded and under-resourced West Side, had reinforced and accelerated the widening gap between the city's rich and poor.[50]

This sharp discrepancy, ironically enough, was replicated in the decade-long decline in San Antonio's position relative to the state's other major cities, Dallas and Houston. Despite San Antonio's intense growth during the Roaring Twenties, and notwithstanding the rapid wholesale reconfiguration of its central business district, the city had lost ground compared to its in-state rivals. Using the number and scale of skyscrapers as the metric of urban power and prestige, which as Goldberger argued was how urbanites of the era evaluated their communities' comparative strengths, the Alamo City's new skyline did not measure up. A 1929 statewide survey identified 135 skyscrapers in the Lone Star State, and the *Express* cheered news that 17 of them graced San Antonio's streetscape and that the Smith-Young Tower was the state's second tallest. But Houston outdid San Antonio: its Gulf Building topped out at 37 stories, and the city was home to a total of 25 tall buildings. Dallas won the overall competition with 34 skyscrapers. Embedded in this data was a signal of what the 1930 census would reveal: San Antonio, which a decade earlier had been the state's largest metropolis, had shrunk to third. This demographic comedown paralleled the city's less robust economy, which already had begun to falter in response to the Great Depression.[51]

That long and grinding economic collapse, which halted major construction projects in the city, also put a hold on the dreams of river reformers who for decades had fought to remake the river itself. Their goal to turn the river's Great Bend into a pedestrian promenade could not be effectively realized until the Olmos Dam was constructed and the river pacified. In the late 1920s Robert H. H. Hugman developed a plan for this riverside walkway, whose design elements offered a beguiling and befuddling mix of materials that exoticized the city's Spanish past (in much the same way as some Olmos Park mansions and several architect-designed skyscrapers that rose above the river). Hugman's proposal, dubbed the Shops of Aragon and Romula, in concert with river-facing developments like the open-air Arneson Theater and La Villita, a reconstructed early Spanish-era settlement, was ultimately delayed until a steady stream of New Deal–era federal funding could be secured. Although local funds would be expended on these projects, they were a small part of the mix, not least because of river fatigue on the part of public officials and voters.[52]

The first signs of this exhaustion emerged in the run-up to the 1926 bond election. The city's flood prevention committee initially proposed a massive $5 million issue, $1 million of which would be committed to still-necessary flood control infrastructure and river channelizing. Over the next several months, after pushback from tax-weary residents and city commissioners, the final offering was whittled to $3.6 million with a mere $200,000 committed to flood prevention initiatives. On November 15 the bond passed handily, but taxpayers were getting restless.[53] That prospect seemed to please the San Antonio Light, whose pages hitherto had been a source of uncritical boosterism. Like a growing number of voters and public officials, the newspaper chafed at what was perceived to be the high price of flood protection: "Into the maw of the river has been poured $3,000,000 by the taxpayers of this city since 1924." Noting that mayor C. M. Chambers "has already voiced his impatience at the ceaseless flood prevention program," the Light thought it might be time to tame the river's "ravenous appetite."[54]

Perhaps the most significant signal that the city was tired of the subject of flood control was the forced resignation of Col. S. F. Crecelius. As director of the city's flood prevention efforts, he once could do no wrong. But after Mayor Tobin died in 1927, his successor began to undermine Crecelius's position when the engineer made a serious miscalculation of the required width for a major cutoff channel through the city. Mayor Chambers slashed Crecelius's salary by 40 percent, and when the engineer did not take the hint, in February 1928 the mayor closed the entire flood prevention office. With Crecelius gone, Chambers reopened the office for the next year before delivering the final coup de grâce, ousting twenty-nine of its thirty employees. By 1929 the city's commitment to flood control—and its associated costs—was over. It would take another twenty-five years and a new set of political actors and agencies to emerge before flood control again became a critical factor in local civic affairs.[55]

┤ SIX ├

UPRISING

Zarzamora Creek is an eight-mile dusty channel that falls away from low hills, part of the Balcones Escarpment delineating the northwestern rim of San Antonio, high ground that is now the site of the San Antonio Medical Center. The creek twists and turns in a southwesterly direction, then breaks south and east before flowing into a retention pond known as Elmendorf Lake, merging there with Apache Creek. Their combined stream flow then pushes southeast and joins with Alazán Creek before the final confluence with the San Pedro. A mile or so later, the flow of these four creeks merges with the San Antonio River. Zarzamora Creek is thus a small part of the San Antonio River's 4,180-square-mile drainage system.

Perhaps the creek's frequent dryness and small size are why the Spanish, who named it, neglected to demarcate it in their eighteenth-century maps of this sprawling watershed, a watershed that made arable the river valley where they sited five religious missions, built a civilian community and a presidio, and irrigated innumerable farms and ranches. For these settler-colonists, like most modern-day San Antonians, Zarzamora Creek left an almost invisible mark on the land.

Yet the creek has had a nasty habit of becoming quite noticeable when it rains. Like the other West Side creeks, its flow increases quickly as rainwater cascades down the slopes and ravines that serve as its tributaries. Water rises fast within its narrow confines, accelerating rapidly, so that if the creek leaps out of its banks it will ram into the built landscape with considerable power. Flash floods are aptly named when it comes to the Zarzamora.

One such flood forced Zarzamora Creek into public consciousness. On August 8, 1974, a torrential downburst erupted over San Antonio as a result of a weak cold front that moved south and west across the Edwards Plateau and collided with thick, moist, and rising air streaming north from the Gulf of Mexico. As often happens on such occasions, the creek immediately became a torrent and rolled out of its banks, inundating the low-lying neighborhoods along its course. The floodwaters flushed more than forty families from their homes, swept away a bridge spanning the creek, and swamped nearby streets. Once the fetid waters receded, they left behind a tangled mass of old tires, broken bottles, dented tin cans, sodden mattresses, tree limbs, and grass-choked shopping carts.[1]

Such damage was normal on the West Side barrio, whose environmental stressors and public health concerns seemed as integral to its social ecology as the creek was to its physical geography. Few in this community felt they had the power to alter what appeared to be natural; few believed they could make themselves heard in city hall. Their sense of powerlessness was a logical and deliberate outcome of a community under the control of a closed, Anglo-dominated power elite. "San Antonio was run by a fairly well-knit oligarchy of well-off, very well-off people," organizer Ernesto (Ernie) Cortés asserted. "All of whom, mostly whites and WASPS, lived in the northeast quadrant of the city and kind of dominated the politics and economics of the city." Because they did not spend "a whole lot of time and energy and attention or resources . . . on the older areas of the city," they ignored those citizens' plight—as had happened in the aftermath of the 1921 flood—whenever dark clouds blew up on the horizon.[2]

The 1974 storm was different, however. The hard-driving rain and its filthy debris provoked a public outcry that could not be contained in the narrow confines that had long defined San Antonio's segregated civic arena. Making ready use of this storm was a handful of grassroots activists. Over the previous year, and under the banner of Communities Organized for Public Service (COPS), they had been quietly organizing among community activists, union organizers, and Catholic parishes on the city's West Side. Ernie Cortés, who had trained at the Saul Alinsky–founded Industrial Areas Foundation, was its initial galvanizing presence. He estimated later that he held more than a thousand face-to-face, one-on-one interviews with potential allies, paying special attention to members of the local machinist union and the American Federation of Government Employees. He also interviewed pastors about key figures in their parishes, met with women who worked in and outside the home, and discovered along the way what issues West Siders believed were paramount in their daily lives. As he listened, Cortés sensed who in the Latino population would be willing to go public with their collective grievances.[3] Some of the pivotal problems were the twinned issues of drainage and flood control, so when the 1974 flood raced through the community, West Side residents turned the flood's damaging impact to their benefit, building momentum for a well-planned assault on discriminatory public policies that routinely forced the poor, disadvantaged, and under-resourced to scramble for higher ground, climb onto a roof, or cling to the nearest tree as rampaging waters submerged their neighborhoods.

Wielding what we now refer to as the language of environmental justice, these protestors challenged the social inequities, political discrimination, and disproportionate burdens they endured, and did so through an intense process of self-education that laid bare the political context of their physical world. This in turn allowed them to identify the means they could employ to better protect their lives and livelihoods, make their community less flood prone and more habitable, and, along the way, wrest power from a white elite whose privilege rarely had been contested. Theirs was a classic example of what theorist Antonio Gramsci calls a "subaltern" struggle, a pitched effort to break from "dominant political formations" to achieve autonomy, however limited, and to press their own claims on the public arena and political affairs.[4]

Laura Pulido uses Gramsci's insights about how "innovatory forces" can emerge to challenge hegemonic authority to describe the advent of a "new form of environmentalism" embedded within the Chicano movement in the Southwest. This environmentalism, she writes, is "counter-hegemonic" because it exists "in opposition to prevailing powers." The emergence of such "explicitly oppositional" organizations depends in part on their ability to alter "the perceptions and meaning of environmental problems and actions on the part of both observers and actors, as a subordinated group's responses to continued economic and political marginalization, struggles over identity, concerns for quality of life, and the continued degradation of the physical environment." The ultimate goal, she notes in language that dovetails with COPS's aspirations, has been "to change the distribution of power and resources to benefit the less powerful."[5]

Strikingly, COPS's subaltern resistance has not been incorporated into scholarly analyses of the origins and significance of the environmental justice movement in the United States. Historians have focused almost exclusively on a later set of events occurring in the late 1970s and early 1980s in Warren County, North Carolina, where a grassroots organization emerged to rally citizens—mostly African American—against the county's and state's decision to locate a toxic-waste landfill in their community. After the Tar Heel State announced in 1978 without seeking public comment that roughly forty thousand cubic yards of polychlorinated biphenyl–contaminated material would be trucked to former farmlands the state had purchased in Warren County, residents fought back in the courts. Four years later, when their legal challenges concerning the dump's potential to compromise local groundwater and hinder other economic development failed, they tried to disrupt the transfer of the contaminated soil. Making shrewd use of local, state, and national media; gaining the attention of civil rights leaders such as Benjamin Chavis, then the head of the United Church of Christ's Commission for Racial Justice; and demonstrating a willingness to go to jail for the cause (there were more than five hundred arrests), activists devised what is often thought to be a new form of environmental protest. By locating the toxic dump in the rural and predominately African American county, the state, they alleged, was practicing "environmental racism," a term Chavis is credited with coining. "The community was politically and economically unempowered," one resident declared. "That was the reason for the siting. They took advantage of poor people of color."[6]

This melding of concerns about social inequality and environmental quality has had a profound impact on the political arena. The emerging environmental justice movement sparked a new generation of civil rights activism and provoked mainstream environmental organizations, such as the Sierra Club and the National Resources Defense Council, to widen their hitherto almost exclusive focus on wilderness protection to include attention to degraded human habitats. The movement has also compelled local, state, and federal governmental agencies to better account for the risks they often impose on disenfranchised people and disempowered communities and to open up their decision-making processes for public examination. These transformative responses, Eileen McGurty argues, are what make Warren County so central to the defining narrative of the emergence of environmental justice as a movement: "The Warren County events ... were significant in the crystallization of environmental justice in three ways: opponents of the hazardous waste landfill were arrested for civil disobedience, people of color were involved in a disruptive collective action against environmental regulatory agencies, and national-level civil rights activists supported an environmental issue through disruptive collective action." Comparing these consequences to those that emerged in the aftermath of the bruising early-twentieth-century battle over the damming of Hetch Hetchy Valley in Yosemite National Park, McGurty notes how each moment helped "articulate [a] common purpose, create a mechanism for building solidarity among movement participants, and allow for the development of ongoing strategic collective action." Even though the uprising in North Carolina failed to stop the construction of the landfill, McGurty argues that the "importance of Warren County for the development of the movement cannot be overemphasized."[7]

Of similar significance were COPS's protests in the early 1970s over the inequities of flooding in San Antonio. Drawing on the heritage of Latino political activism that had emerged before World War II and intensified thereafter, the new organization's potent challenge to the city's willful neglect of low-lying, poor neighborhoods gave voice to the Latino citizenry and morphed into a full-scale assault on an array of related environmental, social, and public health issues that culminated in the re-

Candidates for District 6 during an "accountability session" held by COPS in 1983.

writing of the city's governing charter. San Antonio politics were altered significantly, and as a result COPS's organizing tactics spread to other urban centers with large Latino populations; the organization's energetic activism helped change the political dynamics of the Southwest.[8]

Yet COPS had other and older antecedents. One of the earliest and hitherto unacknowledged of them was Cruz Azul Mexicana (Blue Cross). Like COPS, it drew its members from the local Mexican American population. Like COPS, its membership—leaders and led—was predominantly women. Also like COPS, it brought to bear local knowledge, expertise, heritage, and culture to define its activism. Most tellingly, like COPS, it found its feet in and proved its worth during a flood, in this case the disastrous flood of 1921.

The local chapter of Cruz Azul, one of the first in the United States, may have been established in 1918, though there is some debate about its precise origins. The *San Antonio Express*, after interviews with Cruz Azul's leaders, reported it was established in 1918; *La Prensa*, the local Spanish-language newspaper, noted that the organization was formally founded in November 1921, *after* the flood. As uncertain as the timing of the group's creation is just how closely affiliated it was with the organization of the same name in Mexico. The local chapter's guiding motto, "Charity, Abnegation, and Patriotism," was in line with the larger organization's mission to provide philanthropic support to those in need, and the local chapter was overseen by the Mexican Consulate in San Antonio, the source of a significant donation to Cruz Azul's flood relief work.[9]

Even though it was centrally focused on the needs of the city's West Side, this "grupo de senoritas"[10] expanded its charitable outreach far beyond its home ground, much as COPS would do fifty years later. As the organization worked nonstop to alleviate the suffering of those who survived the 1921 flood in San Antonio, for example, Cruz Azul also sent workers to do the same in San Marcos.[11] It remained active in rural communities across Texas, among them Castroville, Luling, and Asherton, and it even developed an East Texas affiliate in Malakoff. After the San Antonio chapter hosted the first binational conference of Cruz Azul affiliates in 1922, in the succeeding years women-dominated sister chapters proliferated across the country. This striking development has led Julie Leininger Pycior to suggest that "Cruz Azul Mexicana may well have been the first nationwide Latina organization in the history of the United States."[12]

The local chapter's expansive activism was consistent with that of other Mexican heritage mutual aid societies that had been operating in San Antonio since the late nineteenth century. Mutual aid societies formed, as elsewhere, to assist and better the quality of life of newcomers to the city. Indeed the city was at the epicenter of this development, Pycior notes, emerging as the "unofficial capital of the *mutualista* movement. Not only did more Mexicans settle there than in any other city, but it also served as the railroad hub for immigrants heading elsewhere." This flow of people, the existence of social networks, and communal activism, along with the considerable needs of the local Mexican / Mexican American community, meant that by the early 1920s, "countless *mutualista* groups were operating in San Antonio, some dating back to the 1870s and 1880s."[13] Each of them, according to *La Prensa*, arguably the most important Spanish-language newspaper in the United States, was devoted to "the betterment and good name of *mexicanos*."[14]

Among the services these agencies provided to their members in Texas were funeral insurance and educational opportunities, model labor contracts, and legal and medical aid. They also sponsored cultural and patriotic events that may have offered attendees momentary respite from the intense social and economic discrimination that pervaded the Lone Star State. One critical consequence of the prejudicial environment—physical and political—Mexican Americans inhabited in San Antonio was that they were essentially locked into the least-skilled and lowest-paying occupations and compelled to live as renters in some of its most destitute neighborhoods. These were the very people and places most victimized by the 1921 flood.[15]

These were also the people and neighborhoods Cruz Azul began to serve in the early morning on Saturday, September 10, hours in advance of the Red Cross.[16] Cruz Azul members set up two makeshift relief stations, the headquarters of which was located at San Marcos and San Fernando Streets. Sited "a few yards from the Alazán, on a spot where the water rose 20 feet during the flood," the *Express* reported, the station was sheltered "on the west side by a pile of debris which the receding waters of the flood left, and the kitchen and office are side by side out of doors." Here at the flood's epicenter, Cruz Azul's volunteers mimicked the Red Cross's branding by hanging two large banners with white lettering on a blue background declaring: "Cruz Azul Mexicana—Comida y Ropas Gratis" and "Se Imparte Comida Aqui."[17]

The organization's open-air office, consisting of a single table, chair, and typewriter, was set on a level slab that formerly had been a wall of a flood-ravaged building that had washed up on the site. Staff also placed long benches and tables under trees and built a small, wired-off enclosure for the kitchen. Many of the women, including Cruz Azul's president, Marta M. De Acosta,[18] and Delfina Tafolla, its vice president, fired up the stoves and proceeded to do "the actual work of cooking and serving the people." The fact that the *Express* reporter stressed that the group's leader did the "actual" cooking reveals one of the distinguishing features of Cruz Azul's relief efforts. Its leaders did not stand on ceremony but rolled up their sleeves and got to work organizing, cooking, and cleaning, actions that testified to their commitments as members of the community they served. Without diminishing the laudable contributions of the US Army and the American Red Cross to West Side residents, Cruz Azul's grassroots efforts had the distinct advantage of being homegrown and hands-on.[19]

To feed and clothe the displaced, Cruz Azul initially gathered much-needed supplies from West Side merchants and storekeepers, as well as from fraternal mutualista organizations. Joaquin C. Martínez, "on behalf of the Masonic Lodges of the ancient Scottish Rite and accepted under the auspices of the Grand Lodge of the State of Nuevo Leon in Monterrey Mexico, offered his services, and of course, brought with him a large amount of clothing and footwear to be distributed among the poor." In addition to his personal donation to the organization, Mexican consul Angel Casarin Jr. conveyed a check for $2,000 from the Mexican government. Part of the money was initially offered to the Red Cross, but "because it will accept no outside aid" Casarin anticipated donating the entire amount to the local mutualista. "The Blue Cross of course can feel no hesitancy in accepting the donation from the Mexican government, as it is a Mexican organization." The consul promised additional funds "if it is shown that more money will be needed to aid our nationals here who have suffered in the flood."[20] The money, like the donated food, enabled Cruz Azul's kitchens to offer a daily menu that consisted of "soup and beans, meat, bread, coffee, and milk." The organization's impact "gradually expanded as word of it was circulated and calls for relief" accelerated, the *San Antonio Light* observed: "Tuesday 100 were fed, Wednesday 175, and Thursday the number totaled 300." Other badly needed support came from a rotating roster of physicians who staffed this location "to give medical attention to the flood sufferers where it is needed."[21]

Cruz Azul offered similar services at its second relief station. Constructed on a flood-wracked property bounded by Leal, Pinto, and Trinidad Streets, it dispensed "food and clothing to the less fortunate of their people," as well as medical aid. Uniquely, it had an ambulance that one account notes had been "responsible for making visits throughout the day to homes and families that live between the El Paso and Laredo streets, distributing aid of all kinds: clothing, food, and medicine." There was never enough to go around, though, especially because of the acute need for clothing at this site; the surrounding neighborhood was home to more than 100 homeless people. Food began to be in short supply, too. By Thursday, September 15, those standing in line in hopes of a hot meal waited in vain: "the food was so late in coming that most of them quit waiting and went home." It was then that the Cruz Azul and the Red Cross began to collaborate, ensuring a steadier flow of food and clothing to West Side residents, a distribution process that Cruz Azul took over as the Red Cross turned its attention to the refugee camp.[22]

Cruz Azul maintained its relief stations well into November, a persistence that earned the organization praise from the local chapter of the Red Cross and this encomium from the summative report "La tragedia de la inundación de San Antonio": "Under the direction of the altruistic lady, Doña Marta M. de Acosta, the works have been incessant and not a corner remains to date, where the honorable and consoling ray of light of the Mexican Blue Cross has not penetrated."[23]

The organization's work of consolation continued apace in subsequent years: tempered by crisis, Cruz Azul's efforts evolved in direct response to new challenges that West Side residents encountered. Shortly after the flood, the US government began to respond to the postwar economic recession, a spike in nativist sentiment, and Red Scare anxieties by rounding up immigrants and initiating a series of mass expulsions. It did so despite the fact that the federal government had recruited many of these same people to fill the country's wartime labor shortages. "Mutualista associations from Texas to Indiana mobilized to assist the thousands of people being pressured to return to Mexico," Pycior writes, a mobilization that became more essential when Mexican president Álvaro Obregón did not deliver on his promises of financial aid for the detainees. The San Antonio chapter of Cruz Azul was at the forefront of this fight and, in tandem with its peer chapters around the nation, provided money, train tickets, and food to those in need.[24]

With the 1928 election of Herbert Hoover as president, the rise of Ku Klux Klan anti-immigrant fervor, and the subsequent stock market crash in October 1929, Cruz Azul, in partnership with other local mutualista organizations, geared up in response to a new round of deportations. Using the same administrative authorities that his predecessor Calvin Coolidge had deployed to control immigration, the new president ordered the State Department to deny visas to the majority of Mexican applicants and, under the provisions of the Immigration Act of 1882, further restricted "any person unable to take care of him or herself without becoming a public charge."[25] Hoover applied this latter restriction as broadly as possible: "In view of the large amount of unemployment at the time, I concluded directly or indirectly all immigrants were a public charge at the moment—either they themselves were on relief as soon as they landed, or, if they did get jobs, they forced others onto relief." Alberto Remao, a columnist for the West Side newspaper *La Epoca*, scorned Hoover's position: "Naturally this prescription can be interpreted very loosely. Rare has been the immigrant worker who has not in these depressed times been able to become a public charge."[26]

By his actions and rhetoric, Hoover disrupted life in San Antonio's West Side barrios and complicated employment in many of the city's Anglo-dominated businesses. It did not help that city police periodically swept through local parks where laborers gathered in hopes of securing temporary work and "filled the jails with the unemployed," jails that were already getting crowded because the Border Patrol was sending those it had captured crossing the Rio Grande to be incarcerated in San Antonio. The turmoil was so great, the *San Antonio Express* reported, that there was a "hegira of Mexican people who were thrown into a panic by deportations, arrests, rumors."[27]

Although Cruz Azul could do little for those in local jails, it interceded where it could. Much as it had done earlier in the decade, the local chapter once more gathered funds, food, and clothing for those being expelled from the country. In cooperation with other mutualistas, it interceded with civic, state, and federal authorities—as well as with the Mexican consul—seeking to expand the amount of aid, such as railroad passes, available to the deportees; there was also an unsuccessful effort to secure resettlement land grants. Whatever their success, these actions designed to cushion the worst impacts of the federal government's anti-immigration policies testify to the growing sense of solidarity and agency among San Antonio mutualistas, and more broadly within the community they served. This potent form of activism, writes Melita M. Garza, "grew more visible in later Mexican American civil rights actions."[28]

That is one important link between Cruz Azul and COPS, but there is another. With the 1932 election of Franklin D. Roosevelt and the emergence of his New Deal domestic reforms and later Good Neighbor foreign policy, the pressure on the Mexican American community abated. But a new development arose that undercut some of the need for Cruz Azul and other mutualistas. Through a series of initiatives designed to combat unemployment and a host of related social ills—"I see one-third of a nation ill-housed, ill-clad, ill-nourished," the president asserted in January 1937—the Roosevelt administration pushed legislation that created the so-called "alphabet agencies." Among them were the Federal Emergency Relief Administration, Public Works Administration, and National Labor Relations Board. These and an array of subagencies created jobs, rebuilt aging infrastructure, provided food and clothing, and by the end of the decade began constructing public housing to replace condemned slums.[29]

San Antonio was a prime beneficiary of this federal largesse. Construction workers built a new airport, expanded local military bases and airfields, rehabilitated the historic La Villita neighborhood and Spanish-era missions, and developed what would become the River Walk. Whether citizen, resident, or immigrant, many Spanish-surnamed individuals secured work on these and other projects, notably the Alazán-Apache Courts, the first public housing in San Antonio, and by some accounts the first in the nation. Its name is not by happenstance: located close to where the two eponymous creeks had ripped through the West Side in 1921, the housing replaced the dilapidated corrals and jacals that the city had long tolerated. The neighborhood had a high mortality rate, its residents lived in floorless shacks, and its streets were narrow and un-

West Side corrals, 1939.

paved, creating a landscape of disease and death. As journalist Audrey
Granneberg reported in the *Survey Graphic* in 1939: "the West Side is one of
the foulest slum districts in the world. Floorless shacks renting at $2 to
$8 per month are crowded together in crazy fashion on nearly every lot.
They are mostly without plumbing, sewage connections or electric lights.
Open, shallow wells are often situated only a few feet from unsanitary
privies. Streets and sidewalks are unpaved and become slimy mudholes in
rainy weather."[30] Resolving this ghastly situation, like the creation of new
jobs, required more than the voluntary engagements of Cruz Azul and
its mutualista collaborators could generate. It required instead a level of
investment that only the federal government could and would produce.
The New Deal's social programs competed with and marked the decline
in relevance of mutual aid societies like Cruz Azul.[31]

Their waning would open the way for different kinds of activist organizations to emerge on San Antonio's West Side. LULAC (the League
of United Latin American Citizens), which was founded in 1929, distin-
guished itself from the mutualistas and the social services they offered

Map of the 1935 redlining plan, which had an enduring impact on West Side residents' ability to secure mortgages and to build equity.

with its proposal "to integrate the community into the political and social institutions of American life," marking a "fundamental transition from self-help and protection to assimilative activities."[32] A subsequent generation of activists, embodied in the GI Forum (1948), MALDEF (Mexican American Legal Defense and Educational Fund; 1968), and COPS, emerged in the aftermath of World War II and derived considerable clout from their use of new laws such as the Voting Rights Acts (1965) and federal funding streams, like the Model Cities Program (1966), to alter the social, political, and economic situation of West Side residents (and far beyond). Yet even as Ernie Cortés, the architect behind COPS's celebrated rise to power, witnessed firsthand the limitations of *mutualismo*—"growing up, he saw people paying mutual aid dues even as the economic and educational conditions in their neighborhood remained abysmal"—he and his compatriots consciously drew on the legacy of mutualista community-based organizing to drive their modern political activism.[33]

-{ 165 }-

Early on, their energetic engagement was focused on attacking the dire and discriminatory consequences of the management of San Antonio's floodwaters, channeled along ethnic divisions and class lines. COPS knew that since the 1920s, the city's master plans concerning flood control concentrated almost exclusively on the San Antonio River. These documents repeatedly highlighted the construction of the Olmos Dam and praised the downstream widening, deepening, and straightening of critical sections of the river that once had snaked through the city. Its reconstruction had demonstrated its value in May 1935 when a series of powerful storms swept across the metropolitan area. As designed, the Olmos Dam retained the upstream waters, and below it, the newly constructed cutoff channel, a conduit 70 feet wide and 750 feet long that sliced through the urban streetscape, successfully sluiced water out of the central core. The interests of the city's political and economic elite had been secured.[34]

The same could not be said for the West Side. Lacking effective and widespread protective infrastructure, floodwaters cascaded through the barrios. On the evening of May 5, 1935, a late-night downpour of less than two inches sent Martínez Creek "on a rampage," flowing down adjacent streets, inundating homes, and stalling cars caught in low-water crossings; one man was killed when he was hit by an automobile racing through the turbulent water. Four days later a much more powerful storm hammered the city with six inches of rain and hail. Once again Martínez Creek proved especially dangerous: "Often dry, [it] evidently carried eight to ten feet of water when the flood reached its crest." Once more cars were caught in the rushing waters, killing one person. Meanwhile San Pedro Creek, which had received considerable flood control work in the

Olmos Dam during the 1935 flood.

1920s, jumped its concrete-armored banks and morphed into a raging "100-foot-wide river." First responders mounted so many swift-water rescues that Gus Klockenkemper, night chief of the police department, described it as the "busiest night since the flood of September 9, 1921."[35]

The September 1946 flood, powered by seven inches of rain in thirty-six hours, produced the same results. The Olmos Dam held back an estimated five thousand acre-feet of water; below it, newly completed channelization projects captured and diverted an equal amount of water, minimizing damage to the central business district. These measures, however, were of no consequence to the safety of those in the city's western sector. According to a report from the US Geological Survey, shortly after the heaviest rainfall began San Pedro Creek and its tributaries were in full flood. As water poured into adjacent neighborhoods and commercial districts, "flimsy structures were washed down [the] roaring creeks,"

Unpaved streets are inundated with floodwaters in this West Side development, 1946.

nearby stockyards were inundated, and more than a dozen people died or disappeared. Thousands of citizens were left homeless.[36]

Five years later another flood simply reinforced what by then was a longstanding pattern. On June 3, 1951, nearly five inches slashed down in less than two hours. "Olmos Dam, built after the disastrous flood of 1921, was the No. 1 sight-seeing spot as hundreds of cars streamed over the dam to see the acres of impounded water." The West Side offered no such tourist spectacle. The San Fernando Avenue bridge spanning Alazán Creek "went out with the surging floodwaters," as did a wooden bridge that allowed traffic to cross the creek at West Martin Street; another bridge on Zarzamora Street trapped loads of debris and was overtopped but held up. To help his son and three other newsboys make their paper routes, Joe Padilla drove them throughout the West Side, but when they tried to drive over the Cass Street bridge at San Pedro Creek the floodwaters trapped them. No sooner had they abandoned the car and sloshed to dry land than the vehicle was submerged.[37] Like the seven hundred other car owners who had to flee their foundering automobiles, theirs had been a narrow escape.

Local streets and alleys did not stand up to the stormwater's intensity. Across the West Side, "tons of gravel washed from unpaved streets and gaping holes were eaten into pavement." The debris load—silt, rocks, trees, cars, and other material—did its usual damage. Although the caption of an *Express* photograph featuring Janie Rubio, Yolanda Garcia, and David Rubio playing in the swirling waters of the "normally docile Martínez Creek" that inundated their home on Leal Street made light of the situation—"RIVER, STAY AWAY from my door might be the plea of these residents"—the presumption was not funny. The interiors of more than 150 homes were coated in mud and effluent and their inhabitants were forced to evacuate, a predictable result of "San Antonio's old trouble spots, the Alazán-Apache, Martínez and San Pedro creeks," each of which was "roaring with bank-busting loads of water."[38]

That these waterways were known to be so perilous, that in this inundation three more San Antonians perished, and that such levels of death and devastation in the face of flooding were an expected outcome reflected the longstanding disregard for those living on the West Side. This time, however, the flood produced a concerted political response, the first organized effort since the 1920s. The next day, for example, street commissioner Sam Bell Steves urged a $9 million bond issue for more storm sewers, infrastructure that would be targeted at downtown flooding mostly, as the then-current sewers had been unable to handle the storm's fast-moving and large volume of water. "There is no use trying to put the streets into good condition if they are going to be washed away." He had a point, but even more substantive was Bexar County's decision to institute a 30-cent ad valorem tax, half of which would be dedicated to financing nonfederal costs of local flood control projects. Overseeing these initiatives would be the San Antonio River Authority, or SARA. Founded in 1937 as the San Antonio River Canal and Conservancy District, its original purpose had been to plan for a 150-mile barge canal to

the Gulf. Ten years later the US Army Corps of Engineers determined
that the canal would not carry water—its projected costs were inordinate
and its predicted returns minuscule. So, in conjunction with the devastating aftermath of the 1946 flood, the agency dropped all navigation work
on the river in favor of flood control. SARA (it assumed its new name in
1953) would serve as the local sponsor of projects that the Corps of Engineers had recommended in 1951, recommendations that included extensive work on thirty-one miles of the San Antonio River and its tributaries.
This would be a critical first step in changing the public conversation
around flood control and who benefited from it.

That said, the shift did not happen overnight and would prove incomplete. City planners, for example, were aware that flood control initiatives
since the 1921 flood had largely ignored the flood-prone west. Some of
their insights were highlighted in the 1951 master plan for San Antonio.
Although the bulk of that document's section titled "Flood Control and
Drainage" was concerned with ongoing projects to hasten the movement
of water down the San Antonio River and through the central core as
rapidly as possible, it acknowledged that "little or no provision has ever
been made in residential areas to provide an adequate system of surface
water drainage; and the development of the city streets has been accomplished with little or no thought being given to the effect of surface water
on them." One result of this inattention, and one with which West Side
residents were decidedly familiar, was that "houses have been built in low
areas which in time of heavy rain become flooded and the streets in the
area impassable." This was most visible in the spate of new subdivisions
springing up close to Lackland and Kelly Air Force Bases, where so many
Mexican Americans worked. Their developers were not required to build
storm drains or related infrastructure, leading the plan's authors to conclude that "money spent in building streets where drainage facilities are
not provided is wasted money," a simple enough deduction that would
not become policy for more than twenty years.[39]

An almost equally long delay was in store for addressing flooding of
the West Side creeks, which is not to say that there were no flood control projects being planned, constructed, and dedicated—just that they

invariably happened elsewhere in the metropolitan area. The actions of three players dovetailed and ensured this result—the US Army Corps of Engineers, SARA, and congressman Paul Kilday.

In 1954, after almost a decade-long study of the city's manifold flood problems, the Army Corps of Engineers, under the authority of the Flood Control Act of that year, determined "that a serious flood problem exists within the city of San Antonio, an important military center and distribution point for a vast area in southwest Texas, and that a flood protection project for this city to eliminate the flood menace is economically justified." Underlining this justification was the organization's recommendation "that a channel improvement project in San Antonio, Texas be authorized at this time for construction by the Federal Government, substantially as outlined in this report, at an estimated first cost to the United States of $12,906,900." With additional funding provided by the county, for a combined total of $25 million, the Corps's gist is clear: significant relief from flooding would be forthcoming.[40]

Where that money would be spent would be manifest at the December 1957 groundbreaking ceremonies. Noting that "this was the first river project that Bexar County ever attempted where federal contributions had been sought or accepted," Representative Kilday, who had lobbied hard to secure the relevant authorization in the Flood Control Act that encumbered the funding for his district, cheered what he expected to be the end result of the San Antonio River Channel Improvement Project: "When the work has been accomplished and the $15,000,000 supplied by the federal government and the $10,000,000 provided by the San Antonio River Authority, San Antonio will be freed from the constant danger of floods which have caused millions of dollars of property damage and taken many lives."[41] With that, Kilday gave a signal to the operator of the twenty-one-cubic-yard earth remover, who drove the lumbering machine down a ramp into the riverbed to scoop out the project's first load.

The initial phase was focused on a five-mile section of the San Antonio River that would be deepened, widened, and concretized, from Berg's Mill north to the confluence with San Pedro Creek. Once completed, the project would move north from the creek into downtown. By the 1960s the focus shifted again, this time to the far eastern and lightly populated portions of Bexar County and the San Antonio River's drainage system. With the shift in focus went the federal and county funding, and for much of the subsequent decade SARA built a series of dams and channels to control the Salado, Calaveras, and Escondido Creeks.[42]

As important as these projects were, and as enthusiastic as Kilday had been about their potential, they did not stop flood damage in San Antonio as he had predicted "for all time."[43] On May 18 and 19, 1965, another powerful storm roared through the region and again the West Side bore the brunt. Lake Elmendorf overtopped its dam, and Alazán, Apache, and Martínez Creeks barreled out of their banks and swept more than five hundred people from their homes. As it raced down South Navidad Street, Martínez Creek picked two small frame houses off their cedar post foundations. Sara Corrales escaped in the nick of time as her home hurtled three blocks downstream and crashed into a pile of debris. The floodwaters disintegrated the Rodriguez domicile, and part of its roof finally lodged against a bridge blocks away.[44] Many people found shelter at Bowie Elementary on Arbor Place, a street that the 1921 flood had hammered; others stood on the roofs of their cars, caught in the torrents careening down the creeks; and still more were stranded when the Apache took out the Guadalupe Street wooden bridge. For Antonio Martínez, his wife, and two nieces, furniture was their refuge. They tried to escape through the front door of their Delgado Street house, but the water had risen so fast that the outside pressure forced them back in. Within moments their home's interior was filled with four feet of muddy water: "We climbed on top of clothes closets and prayed the water would not reach the ceiling."[45] There was one confirmed death, and a number of eyewitness reports of an early morning drowning of an adult near Woodlawn Lake and a child on Sabinas Street. Police Chief Bischel dismissed these accounts as rumors: "There couldn't possibly be anyone missing at this time without us knowing about it. We're not looking for anybody in the river."[46] Bischel's assertion may have struck West Siders as another example of public officials' systemic disregard for their well-being.

If this recitation of disrupted lives and devastated streets sounds familiar, if it seems to border on a cut-and-paste template, that is the point. No surprise either that SARA's reaction to the punishment the 1965 flood inflicted on the West Side appeared self-serving. Since 1957 it had concentrated almost all its efforts on the San Antonio River and the eastern creeks, and it heaped praised on its endeavors: "Where our flood control projects have been completed," declared David Brune, SARA's assistant manager, "they held up well." This bold claim came despite the photograph in the *San Antonio Express* that flanked his words showing two SARA cranes swamped by the swollen river at Berg's Mill, idled near the site of the 1957 groundbreaking dedication. This despite what Fred Pfeiffer, SARA engineer (and later its general manager), reported about the damage in locations where the agency had completed its work: "Flooding was severe at the end of the river improvement project in southern Bexar County."[47] What had not occurred, though it had been in the long-term plans authorized by the Flood Control Act of 1954 and for which funds had been encumbered, was excavation of the West Side creeks.

SARA and the Bexar County commissioners acknowledged this inaction one day after the 1965 flood. What they did not admit was that for nearly a decade their failure to invest in flood control on the West Side had left residents to wade through high waters, climb up on wardrobes, and seek safety wherever they could. There was an admission of neglect in the very list of projects that SARA and local politicians now deemed "urgent":

> "Martínez Creek from Alazán Creek to Warner Avenue. Some $1.5 million worth of flood control construction is planned for this two-mile stretch."

> "San Antonio River from Lone Star Boulevard to Hildebrand Avenue. This Project, to cost about $2.2 million, mainly involves deepening of the river through the heart of the city."

> "Apache Creek from Brazos Street to Elmendorf Lake, costing $1 million."

"North fork of Martínez Creek, 4/10 of a mile, costing $204,000."

"East fork of Martínez Creek, 1.6 miles, costing $334,000."

"San Pedro Creek from Camp Street to Myrtle Street, costing $1,170,000."[48]

This same to-do list could have been written after the 1935 flood, or the floods of 1946 and 1951.

Ultimately these initiatives would move forward thanks to the intervention of San Antonio's new congressman, Henry B. González. When Kilday resigned from Congress in 1961, a special election was called and González, then serving in the Texas State Senate as the first Mexican American elected to that body, jumped into the race against John Goode, a self-described "militant conservative." Goode might have won had he not been running against González, but the state senator had earned a wide following during his earlier tenure on the San Antonio City Council and in the state legislature, and as a consequence of his recent if unsuccessful statewide campaigns for governor and US Senate. Outspoken on the evils of segregation, he successfully had undercut its application in San Antonio's parks and facilities and mounted two lengthy filibusters in the state senate when white senators tried to resegregate Texas public schools in defiance of *Brown v. Board of Education*. He challenged their actions in an electrifying speech in 1957, declaring he would speak for those without a voice: "I seek to register the plaintive cry, the hurt feelings, the silent, the dumb protest of the inarticulate."[49] His principled stance, sharp wit, and indefatigable energy led to an easy victory over Goode, and he became the first Hispanic politician to represent Texas in Congress; he would serve in the House for thirty-seven years, help establish the Hispanic Caucus, and become a "potent symbol of the opportunities in state and national politics that would become available to Hispanic Americans."[50]

González's political activism and lengthy tenure would prove crucial. So would the fact that he grew up on the West Side in a very political household (his father was the business manager of *La Prensa*) and had firsthand experience with the city's frequently inundated neighborhoods. These experiences would lead him to alter flood control politics in San Antonio. While rain continued to fall over the city in May 1965, González caught the last flight out of Washington so that the next day he and his staff could mount a "quick investigation" of the flooded areas and then immediately return to the Capitol and offer "direct testimony" to the House Public Works subcommittee. His goal was to secure the committee chairman's sanction of an ad hoc subcommittee's visit to San Antonio, where it would conduct public hearings about the flood and publish the results of its investigations. González's ambition was to cast a bright light on the West Side flooding that local politicians had long kept shrouded.[51]

The strategy worked. Within a week, a three-member House team flew to San Antonio to assess the causes of the flood and to seek remedies. It was chaired by Representative Jim Wright of Texas and included Oklahoma congressman Ed Edmondson and Don H. Clausen of California, who arrived in the city on May 23. González arranged for the delegation to meet with the US Army Corps of Engineers, city officials, and SARA; the visiting representatives also toured some of SARA's ongoing projects and two ground-truthing assessments of the damaged West Side. What emerged from the testimony at the public hearings was that city-led projects were directly connected to those SARA was conducting on the San Antonio River, and its projects were underfunded. As for SARA, its intense focus on the river had absorbed most of the federal and local funding that would have paid for flood control infrastructure on the creeks. Striking was the revelation that SARA was poorly administered, that the county had been lax in its oversight responsibilities, and that what SARA's representatives repeatedly referred to as "bottlenecks" in its operations were in fact extensive cost overruns and monumental delays. The agency's representatives then admitted that SARA had completed only one-third of the thirty-one miles of work it had committed to do in 1957, and although the entire project was supposed to have been completed in 1963, it would take another eight years for SARA to fulfill its original commitments. Perhaps its most damning admission was that the agency had run out of money.[52]

Henry B. González speaks at a celebration for an Alazán Creek flood control project, October 1964.

A frustrated González knew about SARA's troubles, but few others in the city were aware of them. It was to educate the wider community that he had asked for an investigative subcommittee. In its final report, the subcommittee shared González's frustration with the delays and budget shortfalls—and the human toll these exacted from people residing on the West Side. "Those damages which did occur brought suffering and heartache to thousands of human beings. Their anguish cannot be measured by the impersonal statistics of cold economic appraisal. To those families whose possessions and hopes were washed away by the cascading San Pedro, Alazán, Martínez, and Apache Creeks, the completion of the San Antonio River Improvement Project will come too late." To prime the pump, the subcommittee received assurances from the Corps of Engineers that it would funnel more funding to SARA, and González successfully appealed to the House Appropriations Committee to secure those federal dollars.[53]

Henry B. González's political skill and community commitments, his marked ability to leverage federal funding streams for his district, changed one set of dynamics that had governed flood control politics in San Antonio. With money came power; with power came the authority to determine where the money would be spent. This son of the West Side forced those who for years had neglected his community to invest their time, energy, and capital in resolving one of its most hazardous features, a process that continued as long as he was in office.[54] Yet in the fetid aftermath of the 1974 flood, a new coterie of strategists introduced a new set of tactics to contest the city's power elite. COPS derived much of its strength from its door-to-door grassroots activism; it was a bottom-up organization, whereas González worked from the top down. They were outside and he was inside, although González would have disputed that too-easy analysis. "Given that the power to influence decisions that affect our lives is concentrated in the established systems of our government, I felt that I could contribute by participating in that process," he once wrote. "There is a place for those who remain outside these processes, but I felt that I could contribute by influencing policy from the inside. Yet even on the inside I have largely remained an outsider because of my refusal to surrender my independence."[55] What González and COPS shared was their conviction that politics was the lifeblood of democracy; everything was political.

That insight drove COPS's approach to flooding and drainage issues in 1974. The organization's members knew that those residing within the web of West Side creeks had waited patiently for flood control infrastructures. They recognized, because González had brought this to light, that SARA's lengthy delays in fulfilling their commitments to restructure the creeks, consciously or otherwise, mimicked the pattern of neglect that city leadership had practiced since the 1921 flood. That is not how the agency's official history describes its actions, however. Its master narrative, which covers SARA's first fifty years, from 1937 to 1987, is devoid of politics and implies that the organization operated outside the political realm. According to that version of history, SARA's engineers and experts crafted logical plans and adopted neutral policies in their battle against urban flooding. Although the text acknowledges, for instance, that public tax dollars underwrote its projects, it never discusses the democratic push-and-pull that produced those funds, or whether and how specific interests might have shaped the bond packages to determine the priority, siting, need, or timing of its projects. This absence is particularly evident in SARA's recounting of its activities during the explosive 1970s, and in the passive voice it employed to describe its actions:

> SARA's flood control work involved the completion of the channelization of the Alazán and Martínez Creeks. Also, SARA was cooperating with the Urban Renewal Agency and the Model Cities staff of the City of San Antonio to complete local requirements for the Apache Creek flood control improvements. SARA coordinated the project and constructed the Nineteenth Street bridge-dam at Elmendorf Lake, the Urban Renewal Agency acquired the necessary land rights, and the City of San Antonio paid all the costs from funds made available through the Model Cities program of the U.S. Department of Housing and Urban Development.... By 1975 channelization of Alazán Creek was completed, and most of San Pedro and Martínez Creeks were completed. The channelization of Apache Creek was completed in 1976.[56]

COPS chose a different framing for its version of these events, employing a more active narrative voice to reflect its oppositional stance. After all, it was in the early 1970s that the new organization began to exert intense public pressure on city officials, and by extension, on SARA, to fund flood control and drainage projects throughout the West Side. It is no surprise then that its chronology begins with a direct clash of wills that erupted on August 13, 1974, at the first public meeting between COPS and Sam Granata, the city manager. Days earlier, Zarzamora Creek had rushed out of its earthen embankment, yet another wet slap in the face of those too long threatened with flooding, an event that infused the five hundred COPS members who gathered at Kennedy High School with a renewed sense of outrage; it even "bolstered our faith in God," one organizer later laughed.[57]

No one was amused at the time, though, as an unsuspecting Granata discovered when he walked into the school's auditorium. An irate audience "demanded action on drainage improvements in the neighborhoods.

They told the city manager they were not interested in long-winded explanations; he could keep his responses to 'simple yes or no answers.'" To bolster their case, they screened photographs of the longstanding pattern of neighborhood flooding and the havoc it caused. "Scenes like these have been there for years and are still the same," activist Ray Kaiser told Granata. "We have decided to not take it anymore. We have decided to make our problem, your problem." When a flustered Granata confessed he did not have the authority to resolve their concerns—"To give a 'yes' and not be able to deliver would be a fraud. I can't do it"—the rambunctious crowd asked him who had the authority. Told that city council was the appropriate body, COPS immediately asked him to place their demands on the following week's agenda.[58]

What happened at that council session has taken on a legendary cast. Hundreds of COPS supporters jammed city council chambers well beyond capacity, and, a reporter recounted, the crowd "broke all the time-honored rules about how citizens are supposed to address the council on a sign-in basis, one at a time." With a theatrical sense of its emerging clout, the audience rose as one when its chief spokeswoman, Mrs. Héctor Alemán, walked to the podium microphone, and crowded around her as she hammered the council into stunned submission. Telling the representatives that her West Side neighborhood lay within the Zarzamora's floodplain, she related how every time thunder cracked overhead and rain fell, houses, streets, the local park, and even the parish church went underwater. Her voice cracked: "How would you feel getting out of bed in the morning and stepping into a river right in your house?"[59] Like the other speakers, she made a blunt claim on the political process: "We are here to demand action. We don't want excuses." Her supporters roared.[60]

Mrs. Alemán's pitch came with a sharp edge. COPS researchers had burrowed deep into the city's archives to locate the 1945 master plan, which contained provisions to reconstruct the Zarzamora as it swung past Mayberry Street, site of many of the bank-busting floods to which

Alemán had alluded and close to her still-sodden home. Bond monies had been committed at that time, but were never spent. Charles Becker, who had been recently elected mayor, swiveled to ask Granata if Alemán's account was accurate. Told that it was, he queried: "You mean to tell me this project has been on the city's list that long and never received a thin dime?" Yes, Granata said. "Well, that's a damn shame," Becker replied. "How many people does this affect?" When advised that the number was more than forty thousand, the shocked mayor gave city staff four hours to locate the financing for the Mayberry Street project and directed the council and COPS to return later that evening to learn the results. The staff complied. Flood control was suddenly and emphatically on the city's agenda.[61]

The raucous public meeting and the flood that precipitated it offered a resounding answer to Rob Nixon's potent query about environmental justice activism: "How can we turn the long emergencies of slow violence into stories dramatic enough to rouse public sentiment and warrant political intervention?"[62] Tired of being victimized by flash floods and slow violence, COPS and its allies drew on their sophisticated conception of a streetscape environmentalism to power their electrifying campaign to change the conditions of life on the West Side. In this, COPS proved relentless. Within three months, it had cajoled the city council into calling a special bond election that would fund fifteen West Side drainage projects to the tune of $46.8 million; the measure passed easily. Over the next three years, COPS would reorient how city council and staff crafted the city's budget, demanding and receiving commitments to underwrite $100 million for neighborhood improvements that ranged from flood control and street repair to trash pickup and water-and-sewer infrastructure. Within a decade, its organizing efforts had netted an estimated $500 million for local environmental and public health projects. As those robust figures suggest, COPS had become adept at finding untapped federal dollars and new uses for previously committed local funds. COPS also compelled the city and SARA to apply for funding from the US Department of Housing and Urban Development to build the new channels, dams, and culverts that began to dry out the often-soggy West Side.[63]

For all its success on the street level, COPS understood the broader implication of its demands for a more habitable landscape—to secure this goal required a shift in the city's power dynamics. To accomplish this,

it helped launch similar organizations on the largely African American East Side and on the ethnically mixed South Side, a pattern of collaborative citywide activism that was unprecedented in San Antonio and the nation. Within a year of the 1974 city council meeting, COPS had also formed an alliance with North Side (and mostly Anglo) environmentalists to battle a massive housing development over the recharge zone of the Edwards Aquifer, successfully protecting the city's sole source of drinking water. That COPS helped build a broad-based, multiethnic, cross-sector alliance was also crucial to its success and its ability to "sow hope."[64]

These achievements had ramifications for local politics, making San Antonio a more democratic place—a transformative effect that gained even greater force as some COPS members joined with the Mexican American Legal Defense and Educational Fund to force the rewriting of the city's charter. For the preceding twenty years, that governing document had mandated that city council elections be run on an at-large basis. This form of governance had enabled the Good Government League, a North Side political machine, to dominate the political arena. Its dominance was reflected in its distribution of the public dollars; it freely spent tax monies to underwrite the expanding suburbs its supporters resided within and the central business district in which they worked. The US Department of Justice entered the fray after charter-revision forces sued the city in federal court, arguing that the at-large political structure violated provisions of the Voting Rights Act of 1965. The Justice Department compelled the city to submit a revised charter to voters in January 1977 that created ten city council districts, and the plan was narrowly approved with 52 percent of the vote.[65]

Get-out-the-vote campaigns on the West Side were critical to the new charter's passage; its precincts reported nearly 90 percent in support of a more equitable government. Four years later, this newly empowered West Side electorate made one of its sons, Henry Cisneros, mayor of San Antonio. On the campaign trail he had given voice to West Side residents' plight and prospects: "Over the last decade, generations of Mejicanos have worked their fingers and shoulders raw and driven themselves to premature old age in order that their sons and daughters could take their rightful places in society," the thirty-three-year-old thundered. "That leg- $\quad$-{ 181 }- acy of sweat and tears provides a moral imperative that this generation assert itself to fulfill its most sacred dreams.... We have a generation of men and women who are hungry for what they can achieve as only those who have been barred from the table can be hungry."[66]

COPS's fight for environmental justice had schooled its supporters in the need for political power, new identity formation, and the broader and democratic benefits that could flow from their subaltern resistance movement. Some activists began carrying this galvanizing message across the Southwest and the nation. Ernie Cortés, COPS's lead organizer in San Antonio, moved to East Los Angeles to launch a similar set of initiatives. Others, all trained at "the University of COPS," fanned out to organize within Latino neighborhoods in Houston, El Paso, Tucson, and Phoenix, as well as in Chicago, New York, and beyond. Willie Velásquez, another San Antonio activist, founded the Southwest Voter Registration Education Project with the goal of bringing Hispanic and other under-represented communities into the political process. Just as Atlanta was an incubator for the Black civil rights movement, so San Antonio was formative ground in which Latino activists conceived, trained for, and tested their ability to reconstruct regional political life. And just as the civil rights movement found its source in multiple communities' activism, so too did the environmental justice movement; its origins now can be traced back to Warren County, North Carolina, and Bexar County, Texas.[67]

Still, locating when and where a protest movement originated is of little concern to those swept up in its development. Much more crucial is clarifying the interwoven set of problems they feel impelled to resolve. At its tenth-anniversary celebrations, COPS's first president, Andy Sarabia, reminded his audience of what they had confronted: "Ten years ago, city leaders said, 'Leave them alone. They're Mexicans. They can't organize.' Today, we have power, we have our culture, we have our faith, we have our communities, we have our dignity, and we're still Mexicans. They feared the successful revolution from a government of the few by the few to a government of the people by the people and for the people. The significance is that the powerless do not have to stay powerless."[68]

The way this disenfranchised people secured power is significant as well. COPS's members first had to place themselves on the map, to recognize that the historic and ongoing flooding of the Zarzamora and other West Side creeks was a consequence of nature *and* politics, of the physical *and* the political landscape. Only then could they identify the need for a social response to this occasionally deadly environmental force. To achieve that level of identification required that they understand their marginal location in local politics, develop a framework to articulate their community's newfound identity, fight to insert themselves in the center of the civic arena, and, once there, use their authority to transform the physical terrain they called home. This process, at once transformative and reciprocal, legitimized their mobilization around a streetscape environmentalism that made it possible for them to imagine a future that was more just and sustainable—and a good deal less treacherous. Nature, the source of so much death, damage, and disarray on San Antonio's West Side, had become the catalyst of its residents' liberation, a liberation that was itself a long-delayed redemption of those who had been swept away in the 1921 flood.[69]

-{ SEVEN }-

AFTERMATH

It was a remarkable, fertile moment. In November 2010 Mayor Julián Castro, District 5 councilman David Medina, Rod Radle of the Metropolitan Alternative Housing Corporation, and Roberto Rodriguez of the West Side Creeks Restoration Project Oversight Committee, with shears in hand, sliced through a long red ribbon to celebrate the official opening of the Apache/Zarzamora Creeks Linear Park. The first greenway on the city's West Side, the park follows the two creeks from General McMullen Drive in a southeastern direction to West Commerce Street where Zarzamora Creek enters Lake Elmendorf. At eight-tenths of a mile, the fully landscaped, bench-lined asphalt path serves walkers, joggers, and cyclists; for those seeking a heart-rate-elevating stationary workout, there are five fitness locations spaced along the walkway. These recreational opportunities were not the only assets that captured the attention of those at the dedication. Changed, too, one attendee noted, was her perception of the creek, once trash-filled, overgrown, and off-putting. "I was scared to come here," Irene Moreno told a reporter. "Nobody came through here." As cormorants and egrets flew by, she meditated on the transformation: "It feels real calm, like you can come out, sit and relax."[1]

In the ensuing decade, other linear parks were developed or extended along Alazán, Apache, Martínez, San Pedro, and Zarzamora Creeks; each greenway linked with the others into an extensive network of pedestrian paths along the banks of the San Antonio River. In the 1990s the city's Open Space Advisory Board had pushed for these linear parks; Mayor

Howard Peak, for whom the pathway system later would be named, offered critical public and fiscal support.[2] The West Side Creeks Restoration Project Oversight Committee helped make the project a reality. Using as its scope of operations the fourteen miles of West Side creeks that the 1954 San Antonio River Improvement Project had identified as needing flood control infrastructure, the committee conducted community-based research, held public charrettes and policy-planning sessions, and in collaboration with the city and the San Antonio River Authority, helped give this nested system of waterways, once so terrifying because their flood-driven energy, a more welcoming demeanor.[3] How appropriate, then, that in 2019 Zarzamora Creek received a second facelift, this time with SARA spending tens of thousands of dollars to relandscape a stretch of the creek fronting Mayberry Avenue. That is precisely where in 1974 floodwaters had rushed over the banks and flushed hundreds from their homes, leading to COPS's political action debut. One reverberation of that political uprising was these creek-restoration projects themselves. It was as if the mountain laurels and sycamores, the cedar elms, oaks, and crabapples, and the shrubs and grasses that were planted along the creek were an environmental peace offering.[4]

The restoration of the West Side creeks, and the linear parks that thread beside them, are woven into a landscape-scale set of initiatives to reengage with, even renaturalize, the region's larger watersheds. The handprint analogy—with San Antonio as the palm and the five major rivers and creeks as the fingers—once again helps to explain the scope of this reengagement. Start where these riparian systems and their many tributaries begin, with water cutting through the Balcones Escarpment or bubbling up from springs, bogs, and seeps before running southeast on the way to the Gulf of Mexico. In the northern reaches of Bexar County, along the upper extent of Salado, Olmos, and Leon Creeks, among others, check dams have been slotted into the thin, stony soil to slow down floodwaters. Those structures that overlie the Edwards Aquifer recharge zone serve a crucial second purpose: to percolate water into the vast karst aquifer that is the key water supply for the county and city. Even as activists demanded this innovative solution, they collaborated with likeminded groups across the city and sympathetic public officials to place a series of parks and aquifer-protection propositions on the ballot. These authorized the city to issue bonds to purchase property overlying the recharge zone; a series of these bonds have easily passed and have provided money for additional drainage infrastructure, dams, and retention ponds. Not incidentally, some of the subsequent propositions helped underwrite West Side creek restorations. This multisector set of upstream developments is consistent with thinking like a watershed, the recognition that we inhabit natural systems and that whatever we do (or do not do) on high ground will have consequences downstream. This concept is akin to the one Forest Service scientist W. W. Ashe had urged the city to adopt days after the 1921 flood—to no avail.

However belated the adoption, Ashe would have praised another watershed adaptation the county and city initiated after another major flood ravaged the region on October 17 and 18, 1998. This one was the result of a perfect storm—tropical moisture funneling into the region from two weakening eastern Pacific cyclones that collided with a cold front over the Edwards Plateau. In the first seventeen hours of the storm, about ten inches of rain fell in San Antonio, with the final total close to sixteen inches. With hail and tornados hammering the area, thirteen people died. There were innumerable swift-water rescues, almost the entire system of streets flooded, major highways shut down, and hundreds of homes were inundated. Impounded water behind the Olmos Dam rose to within eight feet of its spillway, the second-highest level ever recorded.[5] Downstream, the city's $111 million investment in a three-mile-long drainage tunnel bored under the downtown core, which had been dedicated one year earlier, proved a lifesaver. More than three million gallons of water a minute coursed through the twenty-four-foot-wide structure, leading SARA official Steve Ramsey to place the tunnel's achievement in its historical context: "If we did not have Olmos Dam and the tunnel, downtown San Antonio would have been devastated. It would not have been a pretty sight."[6]

Others could attest to the benefit of those long-term investments, among them historian Maria Watson Pfeiffer and her husband, Fred

Pfeiffer, then SARA general manager. They lived in the King William district home that her great-great-uncle, Clarkson Groos, had built in 1859. At the height of the 1921 flood, water rose close to the second-floor porch. Members of Maria's family, including cousin Fred Groos, who later wrote his memories of that harrowing evening, had climbed on a ladder to a nearby elm and then to the roof, where they camped out until the next morning. On Sunday, October 18, 1998, seventy-seven years later, the Pfeiffers sat on the same second-story porch, drank their morning coffee, and watched the river seethe past their home.[7]

Not everyone was so fortunate: residences throughout the metropolitan region were hit hard. In quick order, the county and city took inventories of these properties that lay within the floodplains, assessed their value, and drew up plans to purchase them from willing sellers. County officials determined that it would require more than $9 million to purchase three hundred of these sodden houses. The county applied for and received a Federal Emergency Management Agency hazard mitigation grant to cover 75 percent of the costs and utilized local tax dollars to cover the rest. The advantages of this approach—which the city has also pursued, spending roughly $4 million on buyouts along Martínez and Zarzamora Creeks alone—is that the county and city were pulling people to safety before they found themselves in harm's way. This was a far cheaper and much more environmentally sensitive tactic than building additional dams, concretizing riverbeds, and armoring streambanks. It gave space to nature while preserving the community's well-being.[8]

This twofold aspiration—increasing riparian resilience as well as human health and safety—found expression in another major project that in many respects reversed the previous seven-decades-long fixation on turning the San Antonio River into a flood control channel. Since 1921, the goal had been to box the river into a hard-walled concrete structure designed to sluice water out of the city as fast as possible. That commitment on the part of the city, county, and SARA essentially turned the river into a storm sewer, much as Los Angeles had done to its eponymous river, Phoenix did to the Salt River watershed, and, farther afield, Seoul, Korea, had done to the Cheonggyecheon, which runs through its heart.[9]

The 1998 flood helped reverse the perception that the San Antonio River must remain a straitjacketed conduit. So did the activism of community organizations, the work of public officials, and a lead grant that then senator Kay Bailey Hutchison secured for the US Army Corps of Engineers. No one missed the irony that the Corps of Engineers, which had spent years pouring concrete into the river to control it, would now help undo what it had done.

Two investments in particular reframed the river and the community's perceptions of it. The first project was focused on treating the river as a river, ecologically restoring its physical structure. Drawing on the science of fluvial geomorphology, the Corps of Engineers' funds were directed at undoing the straightening of the river, re-creating a series of pools, riffles, and runs to bring it back to life without neglecting its continued flood control features; the Corps also restored two remnants of the river's historic serpentine course, increasing its sinuosity. The second project reconnected the river with its historic associations with the five Spanish-era missions that form the San Antonio Missions National Historic Park. As part of this reconnection, Mission San Juan's acequia, a ditch that once had pulled water from the river to irrigate arable lands that flanked it, was reopened. In 2011 water once again was channeled into extant farmlands associated with the mission: "It's wonderful to see the oldest water rights in the state of Texas," Mayor Castro said, with a laugh.[10]

The river's remaking did not end with the river. Concrete was removed from its banks here and there and the earth softened with a palette of native plants, flowers, and trees (twenty thousand of them), much as the West Side creeks would be greened. As the riparian habitat was reinvigorated along the Mission Reach, so was the human: more than fifteen miles of hike-and-bike trails from Mission Espada north to downtown were constructed, along with picnic tables, footbridges, and shade structures.[11]

The multifaceted renovations along the river's Museum Reach were more complicated due to the dense urbanization this section runs through, from downtown north to Brackenridge Park. The urban segment, a 1.3-mile linear path with adjoining pocket parks, weaves north from the downtown core along a lock-and-dam operation that allows passenger barges to motor on the river. The 2.25-mile park segment, from Josephine Street north to Hildebrand Avenue, contains a linear path that connects many of the city's cultural and commercial institutions, ending in Brackenridge Park.[12] The comprehensive projects were expensive, with a staggering price tag of $384.1 million from local and federal sources.

More profound than the final sum has been what those dollars produced. To appreciate their generative role, take an imaginary stroll from south to north. You will be moving uphill, following the topography with the humid, prevailing southeast breeze at your back. Pay attention to what you see, hear, and smell, and to what you feel. To walk a watershed is to recognize our place within it—and that of our predecessors, too: putting foot to ground and moving through space is critical to encountering other moments in time. This valley is an archive, a human and environmental repository that contains memories we might recover, even if only partially.

One of those memories is embedded in your upstream path. You will be reversing the thunderous flow of every flood that has pulsed down the watershed, a reminder that disasters are not destiny. In many respects they are human creations and leave a vital record of the interplay between any number of "culebra de agua" and the built environment in the well-watered San Antonio River valley. Like the Spanish settler-colonists who in the early eighteenth century platted a town in between the San Antonio River and San Pedro Creek, the twentieth-century inheritors of this urban landscape acted as if they had little choice over the flood-prone conditions that shaped their daily lives.

That was never true, as this history has made clear. After each inundation, some spoke out about ways to alter the river and creeks to avoid a replication of the death and damage. They were routinely ignored. One of the consequences of inaction was forgetting, an amnesia that was more consequential for some than for others. In San Antonio, then as now, those who paid a higher price and bore a heavier burden frequently were the community's poor, marginalized, and disempowered, those who inhabited San Antonio's impoverished West Side. Their fate is visible and can be documented at precise points along the footpath.

The Mission Reach, appropriately enough, starts at Mission Espada.

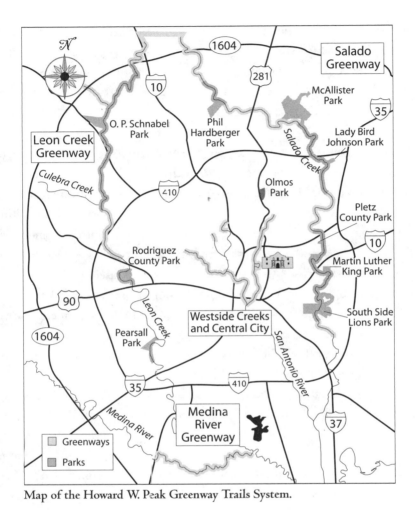

Map of the Howard W. Peak Greenway Trails System.

Head north to Mission San José, which in the past some San Antonians referred to by shorthand as the Second Mission; it was there in September 1921 that Francisca Morales's body came ashore. Continuing north past Mission Concepción, you will soon encounter the King William district: look for the historic Groos homestead, a handy gauge of the 1921 high-water mark. Entering the downtown core and merging with the River Walk, which could only be constructed after the Olmos Dam was erected, remember to gaze at the 1920s skyscrapers looming overhead, and as you stroll under the Saint Mary's Street bridge, recall that Ben Corbo lost his life here as he tried to close down his fruit stand. Farther upstream you will pass Municipal Auditorium, which siphoned $600,000 from the Flood Prevention Fund of 1923, putatively to construct the Romana Street cutoff, an intervention that was not completed until the late 1950s.

At this point, you will be trekking the Museum Reach. Two prominent former breweries anchor this stretch; both suffered considerable damage in 1921, and not far from these important businesses, floodwaters claimed the lives of two children, Ramón and Hortensia Frausto. At Josephine Street take a short detour to walk around the Flood Control Tunnel Inlet Park, where stormwater enters the massive flood control tunnel that was bored under the city. Stop at the San Antonio Museum of Art or the Pearl or farther along at the Witte Museum in Brackenridge Park. Not far from the latter's tree-shaded landscape, hundreds of campers ran for their lives in 1921 as the river swept out of its banks. Beyond, just north of the University of the Incarnate Word, is the frequently retrofitted Olmos Dam—the keystone project for this part of the watershed.

Suppose you instead decide to connect with the West Side creeks trail system at Mission Concepción. That will be easy to accomplish because the mission is sited across from the confluence of San Pedro Creek and the San Antonio River. There you cross an arching pedestrian bridge to the San Pedro greenway where you encounter Confluence Park, an award-winning educational space that teaches the ecological and educational advantages of thinking like a watershed. The 3.2-acre site is on land that the city either purchased or secured via a swap in the mid- to late 1920s. The goal had been to prohibit development on such properties, a passive form of flood control. At some point City Public Service built a storage yard there, but this industrial use has been erased, and a healthier, native plant–rich landscape grows where heavy-duty trucks once parked.[13]

You can continue north along the San Pedro Cultural Park, a four-mile trail that rises in elevation as it comes closer to San Pedro Springs, the heartbeat of the creek. With funding from the county, city, and SARA, the trail is extensively landscaped and contains significant public

Confluence Park's dramatic concrete petals soar above a relandscaped terrain.

art installations. The creek has also undergone rehabilitation. Once a slot canyon–like flood channel, the San Pedro went through a series of reconfigurations starting in the late 1940s. It was rebuilt and rerouted and then moved again with the construction of Interstate 35. In the late 1980s nearly $39 million was spent constructing a six-thousand-foot tunnel diversion project that siphons floodwaters from the creek and carries them underground more than a mile downstream. This initiative made it possible to construct this later pedestrian path, which gives visitors a new way to engage with the creek and with that portion of the city it flows through.[14]

The other option at Confluence Park is to join the Apache greenway that heads northwest from its intersection with San Pedro Creek. As you walk beneath the Mitchell Street bridge it is hard to comprehend how the furious creek swept away the McCray family's parents and six children; somewhere hereabouts all the members of the Morales family, except wife and mother Francisca, also vanished. As you continue along the trail, note the creek's deeper concrete-banked channel, courtesy of long-de-

Children at Confluence Park run their hands over a map of the San Antonio River.

layed funding that also enlarged the capacity of Apache Creek, whose confluence with Alazán Creek soon comes into view. It is not hard to understand why several unidentified bodies came to rest at the Union Stock Yards immediately south of these two creeks' joining, where their raging flow would have briefly flattened and slowed, allowing heavier objects to settle. The Alazán's killing power accounted for the vast majority of those who perished during the lightning-illuminated night in September 1921. Follow the Apache trail until you arrive at Elmendorf Lake and its intersection with the Zarzamora Creek greenway. Farther north you arrive at the spot where in November 2010 Mayor Castro and other celebrants gathered to open the then latest contribution to the city's sixty-nine miles of hike-and-bike trails that hug the San Antonio River watershed.

The ability to walk where so many have died seems paradoxical, not unlike floods themselves; they are destructive as well as creative forces. They can disrupt ecosystems and yet like wildfires can spark new opportunities that give an advantage to rapidly adapting species. It took San Antonio many years to understand that this reciprocal relationship was also true for the human habitat. If there is a silver lining in the city's complicated, distressing history of flood control politics, it is that as it slowly rolled out preventative measures, some residents, activists, planners, and public officials, along with local, state, and federal agencies, began to reimagine the community's relationship with its sprawling watershed. Emerging from these reflections was a more beneficent perception of the many creeks, streams, and rivers that give shape to the topography and geography. In time these agents of change would demand an equitable distribution of flood control infrastructure and would later still dream up a series of linear parks and an accompanying network of amenities that have drawn people back to the river and its tributaries. Floods, it turns out, can nurture vital ecosystems and unleash the human imagination.

-{ 193 }-

─{ APPENDIX }─

NAMES OF THE DEAD, MISSING, AND INJURED

This list of September 1921 flood victims provides their names, ages, and addresses as reported to authorities and represented in coroner's records and local newspapers. The deceased and missing are grouped alphabetically by family name. A list of those who were injured follows. In some cases, ages of these individuals were not listed. In others, their addresses were not known. Often the named streets or alleys here have been wiped out due to construction on the West Side for public housing and other redevelopment initiatives since the late 1930s.

DECEASED AND/OR MISSING

O. E. Brewer, an aviator or stunt flyer from Dallas

Unknown Caravella, 40, missing; 1118 Monterey Street

Felipe Cardenas, 6; Theodora Cardenas, 2; Louisa Cardenas, 15 days; 540 Mitchell Street

Virginia Cavazos, 10; Unknown Cavazos; 1216 Buena Vista Street

Ben Corbo, 42; 1104 Monterey Street

Petra de la Garza, 35; Onesema de la Garza, 4; Carlota de la Garza, 3; Lontardo de la Garza (son of Jesús de la Garza), 1; McAskill Street

Ralph Diaz, 40; Roberta Diaz, 2; 116 Cass Avenue

Alice D. Draeger, 9 months; 2425 S. Flores Street

Rafael Eleasando, 70; Comal Street

Maria M. Escobedo; Josephine M. Escobedo, 7; Jesús M. Escobedo, 4 months; El Paso and Alazan Streets

Guadalupe Escobedo, 38; 317 N. Meldrum Street

Guadalupe B. Falcón, 50; 322 N. Santa Rosa Street

Mr. Juan José Falcón, 60; Mrs. Victoriana Falcón, 60; 750 S. San Marcos Street

Ramon Frausto, 6; Hortencia Frausto, 2; 906 Jones Avenue

Harold Gittinger, 2; 2810 S. Flores Street

Mrs. Amma Gorin, 24; Gilbert Gorin, 9 months; 444 Furnish Avenue

Elena T. Hernandez, 25; Estella Hernandez, 6; Adolfo Hernandez Jr.; 1820 S. Laredo Street

Fred Jones, 6 days; 113 Gertrude Street

M. Lara, 61; address unknown

Ignacia Lopez, 25; Greeven Street; "body found near Pfefferling Stock Yard" [1624 S. San Marcos]

N. A. McCaleb, "about 40 years old"; address unknown; "lost while jitney passenger"

Mr. Tom McCrary, missing; Mrs. McCrary, missing; seven children, missing; south of Mitchell Street

Mr. Morales, missing; Francisca Morales, 36; seven children, missing; near Mitchell Street

Olivia Olivares, 28; 811 Guadalupe Street

Mrs. Rosa Ramirez, 55; El Paso and San Fernando Streets

Mr. Juan Ramón de Zepeda; Mrs. Juanita Ramón de Zepeda, 28; Servia Ramón de Zepeda, 17, missing; Thomasita Ramón de Zepeda, 12; Eusebia Ramón de Zepeda, 11; Estella Ramón de Zepeda, 10; Eduardo Ramón de Zepeda, 6; Donación Ramón de Zepeda, 5; Chano Ramón de Zepeda, 2; Lupita Ramón de Zepeda, 7 months; 1540 S. Laredo Street

Maria Raymond, 12; 1540 S. Laredo Street

Mrs. Rodriguez; address unknown

Saragosa Sandoval, "daughter of Mr. and Mrs. L. Sandoval," 5; 1221
Quincy Street

Andrea Sosa, 14; 540 Mitchell Street

Mr. Francisco Ramirez Soto, 40; 314 Chihuahua Street

Gertrude Southall; 1619 Lakeview Avenue

"Grandma" Vasbinder; Mr. Dave Vasbinder; Mrs. Jeannie Vasbinder,
40; Infant Vasbinder; 124 Cass Avenue

James Ellis West, 11; 705 Tampico Street

Unidentified: a woman found near the Union Stock Yards; a Mexican
boy found near the Union Stock Yards; a white man, about 40, found
near the fairgrounds; a Mexican woman, about 35

INJURED

The sick and injured were treated at Robert B. Green Memorial Hospital,
Santa Rosa Hospital, or P. & S. Hospital Physicians and Surgery. Names
are listed in alphabetical order by family.

Andrew Aganiza; Angelica Aganiza, 25, "fever from exposure";
404 W. Houston Street

Harry L. Bayett, 21; 306 Givens Avenue [last name also spelled as
"Bayette"]

José Bravo, 11; 118 Rankan Alley

J. S. Carl, 522 Agarita Avenue

Jessie Davis, 19; Maggie Davis; 42 Nueva Street

Wynes G. de Rome, "badly injured, spent the night in a tree"; San
Francisco Street

Mrs. R. Diaz, 22; 116 Cass Avenue

Guilbrima Duque, 57; 617 N. Pinto Street

Refugio Elfrieda, 50; 908½ El Paso Street

Bonifacio Esparza, 30; Infant Esparza, 8 months; El Paso and San Marcos Streets

Juan E. Garza, 54; 918 El Paso Street

Miss Josephine Geary, 34; 102 Prospect Street

Mr. Amador Gonzales; 825 S. Richter Street

Concepción Gonzales, 54; Manuel Gonzales, 12; 916 Durango Boulevard

Anita Gutierrez, 43; Francisco Gutierrez, 12; 2425 S. Flores Street. Francisco held his nephew, 5, on his back in a tree on South Flores Street for five hours.

Mrs. J. G. Haddock, 29; Leon Haddock, 7; Miss Bernice Hall, 19, "of Kingsville, visiting the Haddock home"; 213 Ardea Grove

Jennie Johnson, 30, "white"

E. Kidd, "suffering from cut on foot"; 11 West End Lake

Mrs. K. S. Lane, 38; U. L. Lane, 37; 501 E. Glen Avenue

Adelina Mitchell, 17; 211 S. Colorado Street

Albino Navarro, 49; 905 El Paso Street

Mrs. Angela Pena Solis, 22; Lola Solis, 3; Juan Solis, 8 months, 404 Matamoras Street

Mrs. H. Reich, 56; Miss Ethel Reich, 25; 2415 S. Flores Street

A. Rivas, 52, "blood poisoning"; 423 Furnish Avenue

Mrs. Arelia Rochos, 25; Torreón, Mexico

Teodore Solis, "son of Mrs. Joseph Esquizal"; Zapata Street

Mrs. Benita Tudon, "paralyzed"; 108 Alto Alley

Dorotea Treviño, 52; Katherine Treviño, 13; 415 Johnson Street

Incarnación Valciano, 66; Delia, Texas

Mrs. F. Valenzuela, 60; 510 Camaron Street

J. D. Vasbinder; 124 Cass Avenue

-{ NOTES }-

INTRODUCTION: "CULEBRA DE AGUA"

1. "Store Owner Found in River," *San Antonio Express*, September 11, 1921, 2.

2. Ibid.

3. C. E. Ellsworth, "The Floods in Central Texas in September, 1921," Water-Supply Paper 488 (Washington, DC: GPO, 1923), 5.

4. "Laborer Returns to Find Home Washed Away During Flood; Family Missing," *San Antonio Express*, October 14, 1921, 2.

5. C. Terrell Bartlett, "The Flood of September 1921, at San Antonio, Texas," Paper 1485, *Transactions of the American Society of Civil Engineers*, 85: 355.

6. Ellsworth, "Floods in Central Texas," 5.

7. Jonathan Burnett, *Flash Floods in Texas* (College Station: Texas A&M University Press), 28–49.

8. Char Miller, *San Antonio: A Tricentennial History* (Austin: Texas State Historical Association, 2018), 3–11.

9. Fred C. Groos, "Memories of Washington Street, Fall 1987"; in author's possession.

10. "145 Homes Swept Away along Creek," *San Antonio Express*, September 13, 1921, 3.

11. Dulce Watson Bryan to Char Miller, April 15, 1998; in author's possession.

12. Ari Kelman, *A River and Its City: The Nature of Landscape in New Orleans* (Berkeley: University of California Press, 2006); Craig E. Colten, *An Unnatural Metropolis: Wrestling New Orleans from Nature* (Baton Rouge: LSU Press, 2006); Burnett,

Flash Floods in Texas; El Paso, El Paso County, Texas: Letter from the Secretary of the Army Transmitting a Letter from the Chief of Engineers, Department of the Army, Dated February 18, 1965, Submitting a Report, Together with Accompanying Papers and Illustrations, on a Survey of El Paso, El Paso County, Texas, Authorized by the Flood Control Act Approved July 3, 1958 (Washington, DC: GPO, 1965); "Ancient, Historical and Potential Floods," http://repository.azgs.az.gov/sites/default/files/dlio/files/nid1413/tempegi -2-eside2.pdf; Bradford Luckingham, *Phoenix: The History of a Southwestern Metropolis* (Tucson: University of Arizona Press, 1989); Michael F. Logan, *Desert Cities: The Environmental History of Phoenix and Tucson* (Pittsburgh: University of Pittsburgh Press, 2006); Jared Orsi, *Hazardous Metropolis: Flooding and Urban Ecology in Los Angeles* (Berkeley: University of California Press, 2004); Blake Gumprecht, *The Los Angeles River: Its Life, Death, and Possible Rebirth* (Baltimore, MD: Johns Hopkins University Press, 2001), 131–71; John McPhee, *The Control of Nature* (New York: Farrar, Straus and Giroux, 1990), 183–271; Ted Steinberg, *Acts of God: The Unnatural History of Natural Disaster in America* (New York: Oxford University Press, 2000), xix–xx, 201.

13. James L. Slayden, "Some Observations on the Mexican Immigrant," *Annals of the American Academy of Political and Social Science* 93 (January 1921): 121, 125.

14. Laura Hernández-Ehrisman, *Inventing the Fiesta City: Heritage and Carnival in San Antonio* (Albuquerque: University of New Mexico Press, 2008), 56–61; Richard A. Garcia, *The Rise of a Mexican American Middle Class: San Antonio, 1929–1941* (College Station: Texas A&M University Press, 1991); Char Miller, *Deep in the Heart of San Antonio: Land and Life in San Antonio* (San Antonio, TX: Trinity University Press, 2004), 117–27; Rob Nixon, *Slow Violence and the Environmentalism of the Poor* (Cambridge, MA: Harvard University Press, 2013); Steinberg, *Acts of God*, xix–xx, 201.

15. Richard Rothstein, *The Color of Law: The Forgotten History of How Our Government Segregated America* (New York: Liveright, 2017); Christine Drennon, "Explaining Economic Segregation," *San Antonio Express-News*, December 24, 2017, F1, 5; Fernando Centeno, "This City's (Tunnel) Vision," *San Antonio Express-News*, January 14, 2018, F1, 6.

16. "Property Loss Is Heaviest in Downtown Business District," *San Antonio Express*, September 11, 1921, 4.

17. These engineers' evaluations of the Olmos Dam post-construction are housed in the Special Collections at the University of California–Riverside.

PROLOGUE: 1819

1. Martínez (?–1823) became the last Spanish governor of Texas in March 1817 and held that position until 1822; Joaquín de Arredondo (1768–1837) was named the military commander of the eastern division of Provincias Internas in 1813; *The Letters of Antonio Martínez: Last Spanish Governor of Texas, 1817–1822*, translated and edited by Virginia H. Taylor, assisted by Juanita Hammons (Austin: Texas State Library, 1957), 241–43.

2. See Laura Hernández-Ehrisman, "Beyond Adobe Walls: Anglo Perceptions and the Social Realities of San Antonio's 'Mexican Quarter,'" in Margaret Brown Kilik, *The Duchess of Angus* (San Antonio, TX: Trinity University Press, 2020), 250; Rev. Eugene Sugranes, CMF, "Rise of River and Creeks in July 1819 Probably Equaled that of Sept. 10, 1921," *San Antonio Express*, September 18, 1921, 25.

3. Sugranes, "Rise of River and Creeks," 25.

4. Nixon, *Slow Violence*, 8.

CHAPTER I: "DEATH RIDES ON WATERS OF THREE STREAMS"

Epigraph: San Antonio Light, September 10, 1921, 3.

1. *San Antonio Light*, September 10, 1921, 3.

2. Ibid.

3. "145 Homes Swept Away," 3.

4. "Las residencias destrozadas," *La Prensa*, September 13, 1921, 7.

5. "145 Homes Swept Away," 3.

6. Ibid. For a discussion of the US Army's overflights, see chapter 4.

7. "Caught Unaware, Sleeping, Scores Suddenly Swept Off," *San Antonio Express*, September 12, 1921, 5.

8. "La tragedia de la inundación de San Antonio," 2nd ed. (San Antonio: Librería de Quiroga, 1922), 17–22, relates the Zepedas' tragic end; "Mexicans Lose All on the Alazan," *San Antonio Express*, September 12, 1921, 5; "Home from Which Family Flee to Death Untouched by Flood," *San Antonio Evening Express*, September 12, 1921, 1; "Whole Family Dead," *San Antonio Light*, September 10, 1921, 2; "Relatives Seek Bodies in Ruins of Flood Zone," *San Antonio Express*, September 12, 1921, 4.

9. "Home from Which Family Flee"; "Six Flood Victims of One Family Will be Buried in Same Grave," *San Antonio Express*, September 12, 1921, 3; "Parents and Five Children Are Buried in Same Grave," *San Antonio Express*, September 13, 1921, 5.

10. "Six Flood Victims of One Family"; "Parents and Five Children Are Buried"; "Las victimas que han sido sepultadas," *La Prensa*, September 13, 1921, 1.

11. "Flood Extra," *San Antonio Evening News*, September 10, 1921, 1.

12. "La identificacion de otros cadaveres," *La Prensa*, September 15, 1921, 1; Southall remained on the missing list in the English-language newspapers.

13. "Death List, Revised," *San Antonio Express*, September 12, 1921, 1; "San Pedro Creek Tract Dangerous," *San Antonio Express*, September 12, 1921, 5.

14. "Human Features of Flood Tragedy," *San Antonio Express*, September 12, 1921, 5.

15. Ibid.

16. "Powerless to Aid, Scores See Man Slip from a Tree," *San Antonio Express*, September 12, 1921, 4; "Young Girl and Brother Rescue Boys Marooned in Tree," *San Antonio Express*, September 12, 1921, 3.

17. "Powerless to Aid"; "River Rise Brings Widespread Flood," *San Antonio Express*, September 11, 1921, 1.

18. "La tragedia," 26–28; Alfredo Gutierrez's experience is developed in chapter 4.

19. "Siete millones de dolares erdidos," *La Prensa*, September 13, 1921, 1; Cruz Azul's emergency response is detailed in chapter 7.

20. "La tragedia," 36.

21. "Otro victima de la inundación," *La Prensa*, September 14, 1921, 2.

22. "Report Flood Deaths," *San Antonio Light*, October 13, 1921, 7.

23. See chapter 2 for additional details of this campaign.

24. Kenneth Mason, *African Americans and Race Relations in San Antonio, Texas, 1867–1937* (New York: Garland, 1998), 31–38, 268. That this landscape was segregated by race was manifest in a small advertisement in a local newspaper, in which an individual, five days after the flood, offered to exchange a "Ford car to trade for negro lots in West End"; "City Bargains: A.V. Dullye Realty Company," *San Antonio Express*, September 15, 1921, 14.

25. West End Lake is now called Woodlawn Lake.

26. "145 Homes Swept Away," 3; Mason, *African Americans and Race Relations*, 31–38. A small African American neighborhood clustered along Johnson and City Streets

near the west bank of the San Antonio River was destroyed: "Las residencias destrozadas."

27. "40 Known Dead, Fear 250 Perished in Flood That Sweeps San Antonio," *New York Times*, September 11, 1921, 1.

28. The "Death List" that the newspapers published for the first three days after the flood reflected the city's informal recordkeeping: newspapers collected information from the various funeral homes, and the chamber of commerce for some reason vetted that information. What appeared in print was replete with misspelled names, someone anglicized several of the Spanish-language first names (Frank for Francisco, Hortense for Hortencia, etc.), and different accounts contained different addresses for the same person. I have attempted to collate and compare the information reported in the city's three major newspapers, *La Prensa*, the *Express*, and the *Light*; additional biographical material for some of the victims and survivors is contained in "La tragedia de la inundación de San Antonio." The city released its final count (but not a list of names) in the *Light*'s "Report Flood Deaths" on October 13). But its formal record of fifty-one dead is a significant undercount. My research indicates that at least eighty people died, though that figure too is surely not a complete tally, as a result of the chaos of the night and the haphazard process of data collection.

29. "Hungry Children Fed by Red Cross Workers," *San Antonio Express*, September 12, 1921, 3; "Mexicans Lose All."

30. "Una plaga de grillos viene avanzando ha cia San Antonio," *La Prensa*, October 1, 1921, 7; a 2018 study of post-flood insect populations indicates that they are usually decimated, but crickets rebound more quickly due to their ability to recolonize a disturbed environment: Karl A. Roeder, Diane V. Roeder, and Michael Kaspari, "Disturbance Mediates Homogenization of Above and Belowground Invertebrate Communities," *Environmental Entomology* 47, no. 3 (June 2018): 545–50; Joe O'Connell, "Día de las mariposas," *Texas Parks and Wildlife Magazine*, November 2012.

31. "La tragedia," 27–31; "Human Features of Flood Tragedy."

32. The first edition of "La tragedia de la inundación de San Antonio" (Whitt Printing, San Antonio, Texas) was advertised in *La Prensa*, September 18, 1921, 9. The text used here is from the second edition. A third edition is being copublished with *West Side Rising*: see Char Miller, ed., *La tragedia de la inundación de San Antonio* (San Antonio, TX: Trinity University Press, 2021).

33. "La tragedia," 8; in this context, "powerful houses" refers to the major banks, department stores, and other commercial ventures that lined the central business district's east–west arteries; Kenneth Walker, *Climate Politics on the Border: Decolonial Entanglements for Ecological Rhetorics* (Tuscaloosa: University of Alabama Press, 2021), argues that there were multiple authors associated with "La tragedia," journalists who wrote for *La Epoca* and in collaboration with its publisher José Quiroga.

34. "La tragedia," 8, 10.

35. Karen M. O'Neill, *Rivers by Design: State Power and the Origins of U.S. Flood Control* (Durham, NC: Duke University Press, 2006), 80–96, 99–127.

36. "Rainfall in Olmos Friday Night Nearly Double That in City," *San Antonio Light*, September 12, 1921, 3; "Flood Bulletins," *San Antonio Light*, September 11, 1921, 3.

37. "Caught Unaware," 1; "River Avenue Lives Up to Its Name," *San Antonio Express*, September 12, 1921, 2.

38. "Second Child Found," *San Antonio Light*, September 12, 1921, 1; "Majority of Flood Dead Are Buried," *San Antonio Light*, September 12, 1921, 1; there was no information about how Saragosa Sandoval died, though she lived with her parents on Quincy Street north of downtown in a low-lying neighborhood that abutted the San Antonio River (and only blocks from the Frausto family).

39. "City Commission Will Meet Today," *San Antonio Express*, September 11, 1921, 1; "Property Loss Is Heaviest" and "Business as Usual Is Order of Day," *San Antonio Express*, September 12, 1921, 1, 6.

40. "Business as Usual Is Order of Day," 6.

41. *San Antonio Evening News*, Extra Edition, September 10, 1921, 1.

42. "Black Urges Bond Issue to Aid City," *San Antonio Express*, September 12, 1921, 2; "Siete millones de dolares perdidos." City Commission meeting minutes from September 12 to December 8, 1921, are chock-a-block with ordinances detailing which streets were slated for immediate repair, widening, and/or resurfacing and at what cost; thousands of dollars were committed to this single item. Not incidentally, no streets west of San Pedro Creek were included. Municipal Archives, sanantonio.gov/Municipal-Archives-Records/Search-Collections.

43. "Engineer Favors Big Dam across the Olmos Creek," *San Antonio Express*, September 12, 1921, 2.

NOTES TO CHAPTER 2

44. "Repair the Damage—and Protect the City," *San Antonio Express*, September 11, 1921, 4.

45. "Detention Basin on Olmos Vital to Check Floods," *San Antonio Express*, September 18, 1921, 25; the issues of the creeks and the Olmos Dam are extensively discussed in chapter 5.

46. J. B. Gwin, "San Antonio—the Flood City," *Survey* 47, October 8, 1921, 45–66; his work and the Red Cross's efforts in general are detailed in chapter 3.

47. "Paralyzed Mother Held for Hours on Floating Roof by Son," *San Antonio Evening News*, September 12, 1921, 1.

48. William Graham Sumner, *The Forgotten Man and Other Essays*, edited by Albert Galloway Keller (New Haven, CT: Yale University Press, 1919), 476; *What Social Classes Owe to Each Other* (New York: Harper Brothers, 1883); Andrew Carnegie, *The Gospel of Wealth and Other Timely Essays* (New York: Century, 1901).

49. "Loses 25 Houses," *San Antonio Light*, September 12, 1925, 8.

50. Ralph Maitland, "San Antonio: The Shame of Texas," *Forum and Century* 102, no. 2 (August 1939): 51–55; Miller, *Deep in the Heart*, 119–22.

51. "Requiem Mass Heard for Flood Victims," *San Antonio Express*, September 20, 1921, 5; "Mass for Departed," *San Antonio Light*, September 20, 1921, 20.

52. Ramón Frausto genealogy, FamilySearch, https://ancestors.familysearch .org/en/LWL3-5W3/ramon-frausto-1886-1957, accessed May 17, 2020.

53. "Flood Prevention Program Is Begun," *San Antonio Light*, August 26, 1924, 7.

CHAPTER 2: RESCUE MISSION

1. W. Frank Persons to James L. Feiser, September 10, 1921, American Red Cross RG 200, Box 7716: DR 9, Texas San Antonio, Bexar County Flood, September 9, 1921, National Archives and Records Administration (NARA). Unless otherwise noted, all Red Cross communications about the flood are located in this NARA collection.

2. Ibid.

3. James L. Fieser to W. Frank Persons, September 10, 1921.

4. James L. Fieser to W. Frank Persons, September 10, 1921 (see the three telegrams of that date); Gwin, "San Antonio—the Flood City," 46.

5. Marian Moser Jones, *The American Red Cross from Clara Barton to the New Deal* (Baltimore, MD: Johns Hopkins University Press, 2013), 158–57, 173–75.

6. James L. Fieser, "The Red Cross in Disaster," n.p., 1–2; James L. Fieser to Arthur Shaw, September 13, 1921. As important as these photographs were, a "moving picture" would have been better still. In that, the national office would be disappointed that a "flood film" of its work in San Antonio had not been made. But employee Arthur Shaw hoped his supervisors understood why he had been unable to secure footage "as meeting emergency required a superhuman effort." Still, "picture will be mailed by San Antonio Express showing Red Cross Relief." Arthur Shaw to Frank W. Persons, September 13, 1921.

7. *San Antonio Express*, September 11, 1921, 1.

8. *San Antonio Light*, September 11, 1921, 2.

9. *La Prensa*, September 13, 1921, 1; chapter 6 offers a fuller description of Cruz Azul's efforts.

10. *San Antonio Express*, September 12, 1921, 1, 3.

11. Melita M. Garza, *They Came to Toil: Newspaper Representations of Mexicans and Immigrants in the Great Depression* (College Station: Texas A&M University Press, 2018); Julia Kirk Blackwelder, *Women of the Depression: Caste and Culture in San Antonio, 1929–1939* (College Station: Texas A&M University Press, 1984).

12. Natalia Molina, *Fit to Be Citizens? Public Health and Race in Los Angeles, 1879–1939* (Berkeley: University of California Press, 2006), 22–23.

13. *San Antonio Express*, September 12, 1921, 3.

14. Ibid.

15. Garza, *They Came to Toil*, 9–10.

16. Flood Relief Report of the Bexar County Chapter, 4, NARA.

17. *San Antonio Express*, September 15, 1921, 11.

18. *San Antonio Express*, September 13, 1921, 3; September 15, 1921, 11.

19. Flood Relief Report of the Bexar County Chapter, 6.

20. Ibid., 5.

21. *San Antonio Express*, September 13, 1921, 3.

22. Flood Relief Report of the Bexar County Chapter, 5–6.

23. *New York Times*, September 11, 1921, 1; September 13, 1921, 3.

24. *Grand Forks Herald*, September 10, 1921, 1; *Rock Island Argus*, September 10, 1921, 1; *New York Times*, September 11, 1921, 1.

25. *San Antonio Express*, September 12, 1921, 3.

26. "Our New Mayor," *San Antonio Express*, May 10, 1921, 1; "Anti-administration Mayoralty Candidate Wins Close Contest," *San Antonio Evening News*, May 10, 1921, 1; "Death Claims O.B. Black, S.A. Mayor in 1921," *San Antonio Light*, May 7, 1933.

27. See the September 11, 1921, telegram from the General Relief Committee of San Antonio to National Red Cross requesting immediate disbursement of relief funds. Its signers—Sylvan Lang, Claude Birkhead, J. S. Jarratt, C. B. Yandell, A. B. Weakley, Fred K. Terrell, and W. W. McAllister—were members of the city's business elite and key advisers to Black. James Fieser to W. Frank Persons, September 12, 1921, notes that because of the false press reports about the national Red Cross's commitment of relief money, "I assume we must turn over twenty thousand at once to avoid embarrassment to Steves and Red Cross"; other telegrams, many of which crossed over with each other, assert this same need: James Fieser to National Red Cross, September 12, 1921; W. Frank Persons to James L. Fieser, September 12, 1921; W. Frank Persons to Albert Steves, September 11, 1921; and a summary telegram marked "Personal and Confidential," from W. Frank Persons to James L. Fieser, September 13, 1921.

28. James L. Fieser to W. Frank Persons, September 12, 1921.

29. Ibid. In this case, "broadcast" means widely.

30. Ibid.; Persons to Fieser, September 13, 1921.

31. James L. Fieser to Albert Steves, November 4, 1921. Herbert Hoover served as the secretary of commerce under Presidents Warren G. Harding and Calvin Coolidge from 1921 to 1929.

32. "Flood Exaggeration Routed in Circulars Placed with Public," *San Antonio Light*, September 22, 1921, 11.

33. "Realtors Plan Meet," *San Antonio Light*, October 12, 1921, 10.

34. "Lieutenant Governor and Mayors of Texas Cities to Be Guests," *San Antonio Light*, October 16, 1921, 3.

35. "Mayor Cuts Health and Parks Budgets," *San Antonio Light*, October 14, 1921, 1; the city's desire to protect its health tourism business was longstanding. That was one major reason why in 1903 it vigorously resisted the state's implementation of a quarantine banning travel to and from San Antonio during a reported yellow fever outbreak: Ana Luisa Martinez-Catsam, "'Our Local Board of Health Asserts That

No Epidemic of Any Kind Exists in San Antonio': State vs. Local Expertise in the 1903 Yellow Fever Quarantine," *Southwestern Historical Quarterly* 124, no. 1 (July 2020): 1–14; Miller, *Deep in the Heart*, 157–72, probes the longer history of the city's fixation on health tourism and some of its troubling consequences.

36. Catherine Fennelly, *The History of the American National Red Cross* (Washington, DC: American National Red Cross, 1950), vol. 20-B, 8. The American Red Cross was just as surprised in 1926, when, after a hurricane pummeled south Florida, public officials there adopted San Antonio's strategy of refusing to sign on to a Red Cross nationwide donation campaign. "State and municipal officials in Florida decided that the adverse publicity for the state resulting from the Red Cross campaign was so harmful for the tourist trade that they would rather get along without any outside assistance than have more accounts published of distress and suffering," Foster Rhea Dulles, *The American Red Cross: A History* (New York: Harper, 1950), 266–67.

37. Edward Stuart to James L. Fieser, November 28, 1921.

38. James L. Fieser to W. Frank Persons, September 12, 1921.

39. W. Frank Persons to James L. Fieser, September 13, 1921.

40. Edward Stuart to James L. Fieser, November 28, 1921.

41. Arthur Shaw to James L. Fieser, September 12, 1921.

42. Flood Relief Report of the Bexar County Chapter, 7.

43. J. B. Gwin to James L. Fieser, September 30, 1921.

44. Adjusted for inflation, $40 in 1921 equaled $537 in 2021.

45. Gwin, "San Antonio—the Flood City," 46; Henry Baker to James Fieser, October 24, 1921.

46. Albert Steves to James L. Fieser, November 2, 1921.

47. Ibid.

48. *San Antonio Express*, September 12, 1921, 3; describing those fleeing Hurricane Katrina as "refugees" provoked a similar debate over the use of the term to describe those living in the United States: Adeline Masquelier, "Why Katrina Victims Aren't *Refugees*: Musings on a 'Dirty' Word," *American Anthropologist* 108, no. 4 (2006): 735–43.

49. Albert Steves to James Fieser, November 2, 1921. Oakland Street, located just north of the downtown core and running parallel to the San Antonio River, was home to some of the city's white upper- and middle-class residents. By

1929, as part of the straightening of the San Antonio River, Saint Mary's Street was extended north along the route of Oakland Street and the latter's name disappeared from city maps.

50. Two years after the so-called "Anti-Ring" coalition that had originally helped Black get elected in May 1921, the coalition was crushed at the polls. *San Antonio Light*, April 20, 1923, 1, 3; *San Antonio Express*, April 24, 1923, 1, 4; May 9, 1923, 1, 4; Tobin quotation from Mona D. Sizer, *Texas Disasters: Wind, Flood, and Fire* (Lanham, MD: Republic of Texas Press, 2005), 113. Although he was not a candidate, Black tried to influence the conditions of the 1923 election when he sued the commissioners in advance of the May election. Black alleged that they had overstepped their authority in appointing election officers and asserted by contrast that only the mayor held such power. The Court of Civil Appeals of Texas, hearing the case on April 2, 1923, disagreed with Black's argument: Black, Mayor et al. v. Tobin et al., No. 7016, *Southwestern Reporter* 250: 257–58.

CHAPTER 3: MILITARY INTERVENTION

1. *San Antonio Light*, September 10, 1921, 1.

2. *San Antonio Express*, September 12, 1921, 4.

3. Thomas A. Manning, *A History of Military Aviation in San Antonio* (Washington, DC: US Department of Defense, 2000), 9–10; Hiram Bingham, *An Explorer in the Air Service* (New Haven, CT: Yale University Press, 1920), details another source of pilots, the university-based aeronautical schools that he helped develop and supervise; Lewis F. Fisher, *Eyes Right! A Vintage Postcard Profile of San Antonio's Military* (San Antonio, TX: Maverick, 2000), illustrates the material culture that the military's significant presence in San Antonio produced.

4. Dr. P. J. Lipsett, my wife's paternal grandfather, was among the many medical officers trained in San Antonio during the conflict.

5. San Antonio Arsenal, Bexar County Architecture Survey, 1968, https://www.bexar.org/DocumentCenter/View/4102/The-History-of-the-Bexar-County-Courthouse-by-Sylvia-Ann-Santos?bidId=, accessed March 2, 2021.

6. Funston Place is named for Major Gen. Frederick Funston, who commanded the army's Southern Department from Fort Sam Houston and was to have led the US Expeditionary Forces in World War I. But with his death in 1917, that post

went to Gen. John J. Pershing (for whom Pershing Avenue is named). Billy Mitchell Boulevard honors the controversial World War I ace who went on to challenge the leadership of the army and navy for their failure to understand the impact that air power would continue to have in the postwar years. MacArthur Park and MacArthur High School commemorate the legendary, three-war service of Gen. Douglas MacArthur who had grown up in San Antonio and spent part of his career at Fort Sam Houston (an installation that his father had commanded in the late nineteenth century). See David P. Green, *Place Names of San Antonio* (San Antonio, TX: Maverick, 2007).

7. *San Antonio Light*, September 18, 1921, 31.

8. "La tragedia de la inundación de San Antonio," 2nd ed., 36–38.

9. Ibid.

10. *San Antonio Express*, September 12, 1921, 3.

11. Engineer, Eighth Corps Area, USA: "Report on Flood, San Antonio, Texas, September 9–10, 1921," 8. I have deposited this material in 1921 San Antonio Flood Collection (SC.023), Coates Library Special Collections & Archives, Trinity University (San Antonio, Texas). Hereafter "Report on Flood."

12. *San Antonio Light*, September 9, 1921, 1.

13. Ibid., 1, 3.

14. "Report on Flood," 7.

15. Ibid.

16. *San Antonio Light*, September 10, 1921, 1.

17. "Report on Flood," 7.

18. *San Antonio Light*, September 15, 1921, 1.

19. Ibid.

20. Ibid., 8.

21. Ibid.

22. *San Antonio Light*, September 16, 1921, 4.

23. Miller, *San Antonio: A Tricentennial History*, 69–72; Char Miller and Heywood Sanders, "Parks, Politics, and Patronage," in *On the Border: An Environmental History of San Antonio*, edited by Char Miller (San Antonio, TX: Trinity University Press, 2005); Robert L. Lineberry, *Equality and Urban Policy: The Distribution of Municipal Public Services* (Beverly Hills, CA: Sage, 1977), 105–44.

24. *San Antonio Light*, September 16, 1921, 4.

25. Among the many references to this endearing sobriquet are "San Antonio Does Its Part," *Time*, February 21, 1944, 68, and Green Peyton, *San Antonio: City in the Sun* (New York: McGraw-Hill, 1946), 105. Of course, this military-matrimonial reality predated the arrival of the US Army in San Antonio in the 1840s. Spanish and Mexican troops, which had defended the small frontier town from the late seventeenth century to the Texas Revolution in the mid-1830s, also married into local families. Frank de la Teja, *Faces of Béxar: Early San Antonio and Texas* (College Station: Texas A&M University Press, 2016), 93–117.

26. *Trail*, September 16, 1921; *San Antonio Light*, September 18, 1921, 19, reprinted a portion of the special issue, giving its arguments a much wider audience that it might otherwise have received. See also *San Antonio Light*, September 14, 1921, 23, which indicates that copies of the *Trail* would be available on local military bases, hotels, and newsstands.

27. At the battle of Saint-Mihiel, which occurred September 12–15, 1918, the American Expeditionary Force under the command of Gen. John J. Pershing—a well-known figure in San Antonio—launched an assault on the German army near the city of Metz, catching them off-guard and in mid-retreat. It marked an important victory for the US forces.

28. *San Antonio Light*, September 18, 1921, 19.

29. Ibid.

30. Ibid.

31. Information about the army's involvement in the San Antonio flood seemed to have escaped the notice of the Information Group of the US Army Air Service, too. None of the fall 1921 issues of *Air Service News Letter* mention the role of army aviators played in the post-flood analysis. See the relevant issues in https://media.defense.gov/2011/Apr/25/2001330216/-1/-1/0/110425-D-LN615-004.pdf, accessed August 31, 2019. By contrast, *Army Service News Letter*, 5:26, 1–2, offers an extended account of a three-plane expedition to provide aerial reconnaissance of the June 1921 flood that inundated Pueblo, Colorado.

32. Terrence Finnegan, *Shooting the Front: Allied Aerial Reconnaissance and Photographic Interpretation on the Western Front—World War I* (Washington, DC: Center for Strategic Intelligence Research, National Defense Intelligence College, 2006), 3–4.

33. Maurer Maurer, *Aviation in the U.S. Army, 1919–1939* (Washington, DC: Office of Air Force History, United States Air Force, 1989), 131–48.

34. "Report on Flood," 9; less than a year later, Lieutenant Bingham, while
flying from Fort Bliss in El Paso to Denver in August 1922 to give a lecture
on aerophotography, died when his plane crashed near Leadville, Colorado:

earlyaviators.com/ebingham.htm, accessed August 31, 2019; for his part, Lieutenant
Rivers would gain some notoriety in July 1926 when stationed at Luke Field in
Hawaii as head of the photographic section, he flew a high-altitude mission to
photograph the sun's eclipse: *Honolulu Star-Bulletin*, July 9, 1926, 1.

35. "Report on the Flood," 30.

36. Ibid., 30–31.

37. "Lions Told City Now in a Position to Care for Flood Situation," *San
Antonio Light*, September 15, 1921, 7.

38. This was only the latest proposal to divert the Olmos Creek to the Alazán;
as early as 1868, flood control advocates had proposed rechanneling to flush the
Olmos's floodwaters down the West Side creek. William Corner, *San Antonio de
Bexar* (San Antonio, TX: Bainbridge and Corner, 1890), 155. City Commission
Minutes, October 6, 1924.

39. "Engineer Favors Big Dam," 2.

40. "Engineers Expect to Rush Survey of Olmos Basin," *San Antonio Express*,
September 25, 1921, 38.

41. *Guadalupe and San Antonio Rivers: Letter from the Secretary of the Army* (Washington,
DC: GPO, 1954), 8, 32–33, 44, 49.

CHAPTER 4: DAM THE OLMOS!

1. *San Antonio Light*, December 11, 1926, 1, 2; *San Antonio Express*, December 12,
1926; Miller and Sanders, "Parks, Politics, and Patronage"; Lineberry, *Equality and
Urban Policy*, 134.

2. Tobin also indicated that the unspent portion of the $2.2 million bond—
an estimated $600,000—would be dedicated to downstream straightening and
armoring of the San Antonio River and, as possible, clearing away the confluence
of the Alazán and San Pedro Creeks, and that of the San Pedro with the San
Antonio River; *San Antonio Express*, August 30, 1924, 4; *San Antonio Light*, December 11,
1926, 1, 2; *San Antonio Express*, December 12, 1926, 26.

3. *San Antonio Light*, December 11, 1926, 1, 2; *San Antonio Express*, December 12, 1926, 26.

4. "Flood Committee Votes for Concrete Dam," *San Antonio Light*, August 15, 1924, 24.

5. *San Antonio Light*, December 11, 1926, 1, 2; *San Antonio Express*, December 12, 1926, 26.

6. Quoted in Sugranes, "Rise of River and Creeks in July 1819," 25; Metcalf and Eddy, Report to City of San Antonio, Texas upon Flood Prevention, Metcalf and Eddy, Consulting Engineers, 14 Beacon St., Boston, Mass., December 6, 1920, 12–14.

7. Sugranes, "Rise of River and Creeks in July 1819," 25; "Detention Basin on Olmos Vital," 25.

8. Sugranes, "Rise of River and Creeks in July 1819," 25; Lewis F. Fisher, *American Venice: The Epic Story of San Antonio's River* (San Antonio, TX: Maverick, 2015), 18.

9. Quoted in Sugranes, "Rise of River and Creeks in July 1819," 25.

10. Ibid.

11. Ibid.; "Detention Basin on Olmos Vital."

12. William Corner, *San Antonio de Bexar* (San Antonio: Bainbridge and Corner, 1890), 155. The idea of redirecting Olmos Creek down the Alazán resurfaced multiple times across the next century and signals the enduring idea that the West Side and its residents were expected to shoulder the environmental burdens for the wealthier sections of the city.

13. *San Antonio Daily Express*, July 9, 1869, 3.

14. Ibid.

15. Ibid.

16. David Johnson, "'Frugal and Sparing': Interest Groups, Politics, and City Building in San Antonio, 1870–85," in *Urban Texas: Politics and Development*, edited by Char Miller and Heywood T. Sanders, 33–57 (College Station: Texas A&M University Press, 1990); Miller, *San Antonio: A Tricentennial History*, 50, 64, 72, 94.

17. *San Antonio Daily Express*, February 7, 1887, 4.

18. *San Antonio Daily Light*, February 27, 1903, 4; the reference to a flood in 1870 was probably to the July 1869 inundation; *San Antonio Express*, October 3, 1903, 4; *San*

Antonio Light, October 3, 1913, 7, like its rival the *Express,* recaps the major floods to have struck the city and does not mention of the 1819 flood.

19. *San Antonio Express,* October 3, 1913, 1; *San Antonio Light,* October 3, 1913, 1, 2; Amelia Villanueva and her children Olga, Freise, and Amelia were sucked away in front of husband Juan Villanueva; he survived by holding on to a "sapling" for twelve hours: *San Antonio Express,* October 3, 1913, 2.

20. *San Antonio Express,* October 3, 1913, 1.

21. *San Antonio Express,* December 5, 1913, 2, 5–7; December 6, 1913, 1, 3, 6.

22. *San Antonio Express,* October 24, 1914, 1.

23. Ibid., 1.

24. Ibid., 4.

25. *San Antonio Express,* October 3, 1913, 4.

26. *San Antonio Light,* October 4, 1913, 4. Although the *Light*'s worry is understandable, its analogy to Johnstown is not quite accurate. A flood retention dam only holds back water during a flood event; the South Fork Dam was designed to impound the Little Comenaugh River north of Johnstown first to provide water for a canal, then was sold off to private speculators who turned the lake into a private resort (and in the process lowered the dam, among other alterations that made it impossible to release water if needed). David McCullough, *The Johnstown Flood: The Incredible Story behind One of the Most Devastating Disasters America Has Ever Known* (New York: Simon & Schuster, 1968).

27. *San Antonio Express,* October 24, 1914, 5.

28. Ibid., 2.

29. *San Antonio Express,* October 25, 1914, 4.

30. I. Waynne Cox and Cynthia L. Tennis, "Historic Overview and Archival Archaeological Investigation for the San Antonio River Improvement Project: Houston to Lexington," Texas Antiquities Permit 2181, Center for Archaeological Research, University of Texas at San Antonio, Archaeological Survey Report, No. 299, 11–12; Fisher, *American Venice,* 52–53.

31. *San Antonio Light,* September 18, 1921, 29.

32. *San Antonio Express,* September 18, 1921, 25; Cox and Tennis, "Historic Overview," 12; Fisher, *American Venice,* 52–53. I am grateful to Lewis Fisher and Maria Watson Pfeiffer for sharing their insights about the original report, and to Maria for sharing her copy of it.

33. Metcalf and Eddy, Report to City of San Antonio, Texas upon Flood Prevention. Metcalf and Eddy, Consulting Engineers, 14 Beacon St., Boston, Mass., December 6, 1920; *San Antonio Express*, September 18, 1921, 25.

34. *San Antonio Express*, September 18, 1921, 25; Cox and Tennis, "Historic Overview," 11–12; Fisher, *American Venice*, 53.

35. "Would Cut Out Big Downtown Bend in River," *San Antonio Light*, June 18, 1920, 11; "River Project to Be Started at an Early Date," *San Antonio Light*, November 5, 1920, 10; "River Will Have Broader Channel, New Dress Also," *San Antonio Express*, March 13, 1921, 34.

36. Metcalf and Eddy, Report to City of San Antonio, 93.

37. Quoted in "Detention Basin on Olmos Vital." That by their actions civil engineers can reinforce prevailing social discrimination and environmental injustice is revealed in Houston's flood control infrastructure, which, until 2020, was heavily weighted toward wealthier and whiter neighborhoods: Christopher Flavelle, "A Climate Plan in Texas Focuses on Minorities. Not Everyone Likes It," *New York Times*, July 25, 2020, A15. An analogy is the disruptive design and routing of the nation's interstate highway system, beginning in the 1950s. Civil engineers targeted poor and minority neighborhoods for demolition and with the construction of these high-speed interstates severed many of the remaining streets from the rest of the city. A classic example is Interstate 35, running from Laredo through San Antonio, Austin, Fort Worth, and Dallas, all the way to Duluth, Minnesota: it bulldozed through already segregated Latino and African American communities, walling them off and destroying their stability. Tom Lewis, *Divided Highways: Building the Interstate Highways, Transforming Communities*, 2nd ed. (Ithaca, NY: Cornell University Press, 2013); Jane Holtz Kay, *Asphalt Nation* (New York: Crown, 1997); Raymond A. Mohl, "Stop the Road," *Journal of Urban History* 30 (2004): 674–706.

38. Metcalf and Eddy, Report to City of San Antonio, ii.

39. *San Antonio Express*, July 29, 1919, quoted in Cox and Tennis, "Historic Overview," 11; "River Project to Be Started at an Early Date"; "River Will Have Broader Channel."

40. *San Antonio Express*, September 11, 1921, 4.

41. *San Antonio Light*, September 12, 1921, 6.

42. "Las inundaciones en San Antonio," *La Prensa*, September 18, 1921, 9–10; Carl Smith, *The Plan of Chicago: Daniel Burnham and the Remaking of the American City*

(Chicago: University of Chicago Press, 2007); Casey Edward Greene and Shelly Henley Kelly, *Through a Night of Horrors: Voices from the 1900 Galveston Storm* (College Station: Texas A&M University Press, 2000); Amahia Mallea, *A River in the City of Fountains: An Environmental History of Kansas City and the Missouri River* (Lawrence: University Press of Kansas, 2018), 1–32.

43. "Lions Plan Action," *San Antonio Light*, September 13, 1921, 4.

44. An Act for the Control of Floods on the Mississippi River and its Tributaries, and for Other Purposes, 70th Cong., Sess. 1, Ch. 569, 1928, https://www.loc.gov/law/help/statutes-at-large/70th-congress/session-1/c70s1ch569.pdf; Gumprecht, *Los Angeles River*, 176–98; Los Angeles would later secure, after the fatal 1938 flood, millions of federal dollars to concretize the once-rampaging Los Angeles River and to construct dams in almost every canyon in its watershed as well as those feeding the San Gabriel and Santa Ana Rivers: Sarah S. Elkind, *How Local Politics Shape Federal Policy: Business, Power, and the Environment in Twentieth-Century Los Angeles* (Chapel Hill: University of North Carolina Press, 2011), 83–116; Orsi, *Hazardous Metropolis.*

45. "$4,350,000 City Bond Election Is Called for Tuesday, Dec. 4," *San Antonio Express*, October 4, 1923, 1–2.

46. Ibid.; *San Antonio Express*, November 29, 1923, 3; "Think," *San Antonio Express*, November 28, 1923, 1; "Four Engineering Societies Give Support to Bond Issue; Flood Prevention Necessary," *San Antonio Express*, November 28, 1923, 1–2; "Keep San Antonio First," and "Water, Water, Everywhere," editorials, *San Antonio Express*, November 28, 1923, 8; "Straight-Thinking Citizens Will Help San Antonio First," December 1, 1923, Real Estate section, 1, are among a plethora of stories published in the run-up to the bond election.

47. Tobin asserted that a legal challenge to the bond election was "a Ku Klux Klan fight masquerading behind the flood bonds as a smoke screen" ("Klan Bucking Bonds—Tobin," *San Antonio Express*, December 12, 1923, 1).

48. "$4,350,000 City Bond Election Is Called"; second Tobin quote from "Flood Prevention Bond Issue Carries," *San Antonio Express*, December 5, 1923, 1, 4. The naysayers came from a particular demographic—often white middle- and upper-class voters—who lived in a particular area of the city: "Precincts that voted down the proposition in the main were in the heights and some of them

are the same ones that defeated the present city administration in the spring elections." Spared the flood's ravaging power, the well-heeled who lived on higher ground apparently were not fully in favor of protecting those who lived in lower-lying, vulnerable neighborhoods. A subset of the opposition filed and received a temporary restraining order to stop the city from issuing the bonds, alleging the vote was fraudulent and illegal under the Texas constitution, *San Antonio Express*, December 9, 1923, 1. Public officials decried the lawsuit as an attempt, in Mayor Tobin's words, to "gum the works," to delay the sale of bonds. When the restraining order was lifted, anti-bond forces immediately appealed to the Court of Civil Appeals. They lost once again, with the court chiding them for trying to overthrow a legitimate election. Wendover v. Tobin, Mayor et al., No. 7214, April 16, 1924, *Southwestern Reporter* 261: 434–38.

-[217]-

49. "Eight Engineers Would Build Dam," *San Antonio Express*, August 22, 1924, 6.

50. "75 Per Cent of City Bond Money to Be Spent for Labor, Mostly Recruited at Home," *San Antonio Express*, August 24, 1924, 10; "Flood Prevention Program Is Begun," 7.

51. "Flood Prevention Program Is Begun."

52. The following City Commission minutes lay out the purchases of equipment and land acquisitions/swaps/condemnations from the 1923 Bond: September 3, 1924; September 17, 1924; September 29, 1924; October 6, 1924; October 27, 1924; November 3, 1924; November 10, 1924; November 24, 1924; December 10, 1924; January 12, 1925; January 13, 1925; April 6, 1925; July 27, 1925; September 7, 1926. In the 1926 bond, which provided $600,000 for the San Antonio River, and San Pedro and Alazán Creeks, the Alazán received a $17,900 bridge, three low-water crossings totaling $1,700, one land-acquisition for $200, and a land swap. Those were last recorded expenses on it through 1930: City Commission minutes: October 11, 1926; January 27, 1927; February 7, 1927; April 18, 1927; October 17, 1927; November 28, 1927.

53. Seth D. Breeding, "Flood of September 1946 at San Antonio, Tex.," Circular 32 (Washington DC: GPO, 1948), 1–2 (USGS report); "Body of Boy, 3, Found; Flood's Toll at 9 Dead," *San Antonio Express*, October 1, 1946, 20; "Another Survey of Flood-Control Works Is Needed," *San Antonio Express*, September 28, 1946, 2.

54. "Another Survey of Flood-Control Works."

CHAPTER 5: CONSTRUCTION PROJECTS

1. "Finishing Touches Put to Huge Dam," *San Antonio Express*, August 28, 1926, 5.

2. "Do Not Forget the Auditorium," *San Antonio Light*, August 2, 1924, 4.

3. Karen Stothert, *The Archaeology and Early History of the Head of the San Antonio River*, Southern Texas Archaeological Association, Special Publication 5, and Incarnate Word College, Archaeology Series 3.

4. Metcalf and Eddy, Report to City of San Antonio, 13a; *San Antonio Light*, September 12, 1921, 3. The dilemma of utilizing upland rain gauges, however well managed, was that the 1921 flood knocked out electricity so that those maintaining the micro weather stations could not directly report to the city about the amount and rate of rainfall. As noted in chapter 4, the city sent two motorcycle police officers to contact the ranchers, but they were forced to turn back because of the tremendous volume of water roaring through the lower basin of Olmos Creek. The officers raced back to city hall to report the impending disaster.

5. *San Antonio Daily Express*, July 8, 1869, 3.

6. Char Miller, "Where the Buffalo Roamed: Ranching, Agriculture, and the Urban Marketplace," in *On the Border*; Miller, *San Antonio: A Tricentennial History*, 80–91; Paul H. Carlson, *Texas Woollybacks: The Range Sheep and Goat Industry* (College Station: Texas A&M University Press, 1982), 49–50, 193–96; Lisa O'Donnell, "Historical Ecology of the Texas Hill Country," City of Austin, Wildland Conservation Division, Balcones Canyonlands Preserve, January 29, 2019, 22–27; see also "Watershed and Other Related Influences and a Watershed Protection Program," Senate Document No. 12, Separate No. 5 (Washington, DC: GPO, 1933), 361–63; Gayle Brennan Spencer, *Last Farm Standing on Buttermilk Hill* (San Antonio, TX: LBJ CommuniCo, 2010).

7. Char Miller, "Reclamation Project: W. W. Ashe, the Forest Service, and Watershed Stewardship," *Forest History Today*, Spring/Fall 2020, 40–49. "Cities Will Continue to Be Devastated Unless State Acts, Claim," *San Antonio Express*, September 16, 1921, 1–2.

8. "Cities Will Continue to Be Devastated," 1–2.

9. William L. Bray, "The Timber of the Edwards Plateau of Texas: Its Relation to Climate, Water Supply, and Soil" (Washington, DC: US Department of Agriculture, Bureau of Forestry, GPO, 1904), Bulletin No. 49, 21.

10. "Cities Will Continue to Be Devastated," 2; O'Donnell, "Historical Ecology," 34, notes that the Ashe juniper (*Juniperus ashei*), which Texans called cedar and is ubiquitous throughout the upper Olmos Basin and across the Balcones Escarpment and Edwards Plateau, was named for W. W. Ashe.

11. *San Antonio Express*, September 16, 1921, 1–2. Ashe reiterated his insights about flooding and siltation in central Texas in W. W. Ashe, "Financial Limitation in the Employment of Forest Cover in Protecting Reservoirs," US Department of Agriculture, Bulletin No. 1430, August 1926, 24.

12. *San Antonio Express*, September 16, 1921, 2.

13. Ibid., 1–2.

14. C. H. Kearney, A. Y. Walton, and R. B. Huffman to Mayor John W. Tobin, and City Commissioners, March 1, 1927, Letters and Photographs Pertaining to Construction of Olmos Dam, Texas, in Special Collections and University Archives, WRCA G248 E7, University of California-Riverside, https://ucr.primo.exlibrisgroup.com/permalink/01CDL_RIV_INST/1jkonok /alma991003372899704706 (hereafter WRCA-UCR).

15. Unsigned letter to A. J. McKenzie, April 8, 1927, WRCA-UCR.

16. Ibid.

17. Ibid.

18. "Dam Will Be Ready in July," *San Antonio Express*, January 9, 1926, 10.

19. C. H. Kearney, A. Y. Walton, and R. B. Huffman to Mayor John W. Tobin and City Commissioners, March 1, 1927, WRCA-UCR, 1–2.

20. Ibid., 2–3.

21. Ibid., 3. Given the litigious nature of San Antonio's political and civic life, it is not surprising that McKenzie Construction Company would sue the City of San Antonio to recover money it alleged the city had not paid it for work on the dam. See McKenzie Const. v. City of San Antonio, 50 S.W.2d 349 (Tex. Civ. App. 1932).

22. The "Think" columnist for the *San Antonio Express* cheered Colonel Crecelius for saving "750,000 out of a 2,200,00 bond issue for flood prevention work" and redirecting those dollars from the dam's budget to "unkink the river's curves into a straightaway course for rising rivers." It appears instead that Crecelius's flawed design and his and the city's cost-saving measures were responsible for undercutting the dam's safety and that of downstream residents and businesses: *San Antonio Express*, January 12, 1926, 1. Fifty years later, in response

to calls by Rep. Henry B. González to repair the aging dam and worried engineers' concerns about its durability, in the 1970s the San Antonio River Authority (SARA) commissioned a series of inspections to determine the stability and safety of the Olmos Dam. These analyses confirmed every one of the 1927 criticisms (apparently without knowing about these earlier critical assessments). When the structure was finally rebuilt in 1980, the dam was elevated eight feet, the road and the arches were eliminated, a spillway was finally added, the foundation was more tightly anchored to bedrock and the eastern and western bluffs, and additional safety features were installed at the dam's footing, upstream and downstream. It is noteworthy that these latter-day contractors wrestled with "excessive groundwater" flowing across the site, as had the original builders. Driving the need to reconstruct the dam was SARA's conviction that if left unmodified it would not survive another 1921 flood of record and its collapse would have disastrous consequences. See Henry B. González Papers, Box 2004-127/63 [10009704], University of Texas Library; "Ceremonies Marking the Completion of Construction of Olmos Dam Modifications," San Antonio River Authority, December 17, 1980; "Olmos Dam Modifications," *Civil Engineering-ASCE*, June 1982, 54–55; Deborah Weser, "Work Ahead of Schedule on S.A.'s Flood Guard," *San Antonio Light*, June 1, 1980; "Dam's Renovators Blow Its Top," *Engineering News-Record*, 204:5, January 31, 1980, 30–31.

23. Interview with Ruth Dubinsky Kallison, September 13, 1977, Oral History Program, Bexar County Historical Commission, 3, 14: Institute of Texas Texan Cultures, https://digital.utsa.edu/digital/collection/p15125coll4/id/1655. Ruth Kallison's sister-in-law, Frances Rosenthal Kallison, conducted the interview. Nick Kotz, *The Harness Maker's Dream: Nathan Kallison and the Rise of South Texas* (Fort Worth: TCU Press, 2013), offers a compelling analysis of the Kallisons' multigenerational impact on San Antonio.

24. "Buying San Antonio's Hills," *San Antonio Light*, August 5, 1924, 7; see "Where Wild Life Survives," *San Antonio Light*, August 6, 1924, 6, which notes that as wealthy San Antonians pushed farther out into the rolling hills to the north, one consequence would be the disruption of wildlife habitat: "The city is slowly but surely pushing these wild things backward and away as the new homes are built for mile after mile." Another editorial, the first in the series "City Limits and County Lines" (*San Antonio Light*, August 4, 1924, 4), argues that San Antonio and its hinterland now are in closer proximity and economic reciprocity: "It is a good

thing for Uvalde, Hondo, Pearsall, and the scores of other towns in this territory to feel that their city limits includes the Alamo; and it's a good thing for San Antonio to feel that her streets do not end at the line where the jurisdiction of her city commissioners terminates." This outward thrust remains one of the city's key
dilemmas: Miller, *On the Border*, 130–57.

25. "We Are Going to the Suburbs," *San Antonio Light*, August 18, 1924, 6.

26. Adjusted for inflation, $50,000 in 1921 had the purchasing power of more than $730,000 in 2021; the cost of Thorman's house was quickly eclipsed: an Ayers and Ayers–designed home was being constructed for $75,000 (which cost nearly $1.1 million in 2021), *San Antonio Light*, May 5, 1928, 49; "New Dam Inspired H. C. Thorman to Develop Olmos Park," *North San Antonio Times*, December 16, 1976, 16; *San Antonio Express*, December 5, 1926, 37; the following paragraphs partly derive from Char Miller and Heywood T. Sanders, "Olmos Park and the Creation of a Suburban Bastion, 1927–1939," in *Urban Texas: Politics and Development*, 113–27.

27. Advertisements: *San Antonio Express*, November 14, 1926, 39; December 5, 1926, 37; in April 1928, the city let a contract to build a bridge over the San Antonio River so that Hildebrand Avenue could be extended from Broadway Avenue on the east to McCullough Avenue on the west. The newly constructed bridge and extension of Hildebrand, which runs on the southern edge of Olmos Park, further boosted sales in the nascent development: *San Antonio Light*, April 22, 1928, 48.

28. Advertisements: *San Antonio Express*, November 28, 1926, 35, 38; December 5, 1926, 37.

29. Advertisements: *San Antonio Express*, November 14, 1926, 39; November 21, 1926, 33; December 26, 1926, 37; Charles E. Beveridge, Paul Rocheleau, and David Larkin, *Frederick Law Olmsted: Designing the American Landscape* (New York: Rizzoli, 1995); Nichols's work in Kansas City is examined in Kevin Fox Gotham, *Race, Real Estate, and Uneven Development: The Kansas City Experience, 1900–1920* (Albany: State University of New York Press, 2002).

30. Advertisement: *San Antonio Express*, November 21, 1926, 33.

31. Gotham, *Race, Real Estate, and Uneven Development*, argues that J. C. Nichols, whose curvilinear street suburban development in Kansas City influenced Thorman's spatial layout for Olmos Park, was also a pioneer in hyper-restrictive racial covenants that Thorman adopted for his San Antonio development; these covenants presaged the federal government's nationwide sanctioning of

segregation-by-redlining: Rothstein, *Color of Law*; Richard Kluger, *Simple Justice: The History of* Brown v. Board of Education *and Black America's Struggle for Equality* (New York: Alfred Knopf, 1976), 120, 246–47; Thomas Philpott, *The Slum and the Ghetto: Neighborhood Deterioration and Middle Class Reform, Chicago, 1818–1930* (New York: Oxford University Press, 1978), 189–96, 255–56; the enduring quality of these policies, formal and informal, are the subject of Drennon, "Explaining Economic Segregation," 6; Centeno, "This City's (Tunnel) Vision," F1, 6.

32. Warranty Deed, "Olmos Park Estates," Bexar County Deed Records, v. 954, p. 395; Philpott, *Slum and the Ghetto*, 89; Mitchell Place advertisement, *San Antonio Light*, October 1, 1921, 35.

33. Warranty Deed, v. 954, p. 395; *San Antonio Express*, December 5, 1926, 37, 41, and December 26, 1926, 37.

34. *San Antonio Express*, November 21, 1926, 33; Peyton (*San Antonio*, 18, 22) offered this impressions of those calling Olmos Park home: the community's social hauteur attracted a certain clientele, among them "rich oilmen and cattle ranchers," and as a result you would not "find many sheep and goat men living in fine Olmos Park mansions. They do not have the social pretensions of the old cattle families."

35. This and the next paragraph draw on data from the San Antonio City Directory, 1924 through 1934. Thanks to Kay Reamey for her initial research assistance; notes in author's possession.

36. Frances R. Kallison interview with Perry Kallison, July 18, 1976; Kotz, *Harness Maker's Dream*, 124–32, on the Kallison family's life in the affluent enclave of Olmos Park.

37. Paul Goldberger, *The Skyscraper* (New York: Alfred A. Knopf, 1981), 3–4, 8.

38. "Flood Prevention Means $12,000,000 in Building," *San Antonio Light*, August 14, 1924, RE-1; "Removal of Flood Hazard Big Boost for San Antonio," *San Antonio Express*, November 14, 1926, 36.

39. *A Preliminary Report on a System of Major Streets for San Antonio, Texas* (Saint Louis: Harland Bartholomew, 1930), 216; a sampling of overhead newspaper photographs include "San Antonio's $1,000,000 Street Widening and Extension Project," *San Antonio Express*, November 14, 1926, 36; "When the Camera Looks Down on Busy, Sunny San Antonio," *San Antonio Express*, December 12, 1926, 41; "Giant Office, Hotel Structures Evidence of City's Growth, Future," *San Antonio Express*, December

NOTES TO CHAPTER 5

11, 1927, 43; "Downtown Building Programs Total Millions," *San Antonio Express*, August 12, 1928, 36.

40. Looking for bargains: "District Making Good Progress," *San Antonio Express*, February 6, 1927, 33–34.

-[223]-

41. Julius M. Gribou, Robert G. Haney, and Thomas E. Robey, eds., *San Antonio Architecture: Traditions and Visions* (San Antonio, TX: AIA San Antonio, 2007), 82.

42. "Bowen's Island Will Get $500,000 in Buildings," *San Antonio Light*, August 3, 1924, RE-1; "How Competed Plaza Hotel Will Look," December 19, 1926, 48; the $1 million investment added five hundred hotel rooms to the city's inventory; Gribou, Haney, and Robey, *San Antonio Architecture*, 89.

43. Gribou, Haney, and Robey, *San Antonio Architecture*, 90.

44. Ibid., 76; "Giant Office, Hotel Structures Evidence."

45. Gribou, Haney, and Robey, *San Antonio Architecture*, 47–48.

46. Ibid., 78; "25-Story Skyscraper," *San Antonio Express*, October 6, 1929, 62.

47. Gribou, Haney, and Robey, *San Antonio Architecture*, 90.

48. Ibid., 46.

49. "Foundation Laid for 10-Story Structure at Pecan and St. Mary's," *San Antonio Light*, August 21, 1924, 12; Gribou, Haney, and Robey, *San Antonio Architecture*, 43; "Structure to Cost $225,000," *San Antonio Express*, January 16, 1927, 29; the Real Estate, or Hendrick, Building is undergoing restoration to become a boutique hotel: Joshua Fechter, "Blighted Downtown San Antonio Building to Become 47-Room Hotel," *San Antonio Express-News*, March 25, 2019; another centralizing location fronted South Flores Street. Known as the Kallisons' Block (1926), its façade replicated some of the same Spanish Colonial Revival elements found in the family's Olmos Park homes. Inside, the block-long, two-story edifice contained the Kallisons' legendary western store and office space for regional livestock interests: Kotz, *Harness Maker's Dream*, 115–16.

50. Miller, *San Antonio*, 104–16. Depression-era inequities continued to trouble San Antonio into the twenty-first century: Miller, *Deep in the Heart*, 117–27; Mary Jane Appel, *Russell Lee: A Photographer's Life and Legacy* (New York: Liveright, 2020).

51. "San Antonio Has 17 of 135 Skyscrapers in State of Texas," *San Antonio Express*, November 10, 1929, 54; Goldberger, *Skyscraper*, 3–4; Miller, *San Antonio*, 104–55.

52. Miller, *San Antonio*, 104–16; Fisher, *American Venice*, 90–100, 106–24.

53. "Bond Issue Committee Will Meet Wednesday," *San Antonio Express*, June 8, 1926, 6; "Big Improvement Fund Necessary Committee Says," *San Antonio Express*, June 15, 1926, 1, 3; "Wright Confident Bonds Will Carry," *San Antonio Express*, November 15, 1926, 1, 2; "$3,600,000 to Be Made Available on Improvement," *San Antonio Express*, November 16, 1926, 1, 4.

54. "S.A. Pours Millions into River Channel" and "Proposed Bond Election to be Delayed," *San Antonio Light*, October 3, 1929, 15.

55. Fisher, *American Venice*, 81–84; "'Gold' Lines Channel for S.A. River," *San Antonio Light*, October 4, 1929, 21; "Another Assault upon the Beauty of the River," *San Antonio Light*, October 5, 1929, 6.

CHAPTER 6: UPRISING

1. Mary Beth Rogers, *Cold Anger: A Story of Faith and Power Politics* (Denton: University of North Texas Press, 1990), 12–13; this chapter is a much-revised version of Char Miller, "Streetscape Environmentalism: Floods, Social Justice, and Political Power in San Antonio, 1921–1974," *Southwest Historical Quarterly* 114, no. 4 (October 2014): 159–77.

2. Peyton, *San Antonio*; David R. Johnson, John A. Booth, and Richard J. Harris, editors, *The Politics of San Antonio: Community, Progress, and Power* (Lincoln: University of Nebraska Press, 1983), 3–27, 191–212; Rodolfo Rosales, *The Illusion of Inclusion: The Untold Political Story of San Antonio* (Austin: University of Texas Press, 2000); Ernesto Cortés Jr. to Lynnell J. Burkett, May 27, 1994, interview transcript, Institute of Texan Cultures Oral History Collection, 4–5, http://digital.utsa.edu/cdm/ref/collection/p15125coll4/id/275, accessed July 27, 2020.

3. Cortés to Burkett, May 27, 1994, 13–20; Ernesto Cortés Jr. to David Todd and David Weisman, April 12, 2002, Texas Legacy Project, interview transcript, texaslegacy.org/bb/transcripts/cortesernietxt.html, accessed July 27, 2020; Joseph Daniel Sekul, "Communities Organized for Public Service: Citizen Power and Public Policy in San Antonio," in Johnson et al., *Politics of San Antonio*; Carlos Muñoz Jr., "Mexican Americans and the Promise of Democracy: San Antonio Mayoral Elections," in *Big City Politics, Governance, and Fiscal Restraints*, edited by George E. Peterson (Washington, DC: Urban Institute Press, 1994), 105–20; Rosales, *Illusion of Inclusion*, 79–40.

4. Quentin Hoare and Geoffrey Nowell Smith, editors, *Selections from the Prison Notebooks of Antonio Gramsci* (New York: International, 1971), 52–53; with overdue thanks to the late Lucian Marquis of Pitzer College for loaning his copy of Gramsci's writings in fall 1974.

5. Stephen Morton, *Gayatri Spivak: Subalternity and the Critique of Postcolonial Reason* (Malden, MA: Polity, 2007), 96–97; James Scott, *Weapons of the Weak: Everyday Forms of Peasant Resistance* (New Haven, CT: Yale University Press, 1985); quotations from Laura Pulido, *Environmentalism and Economic Justice: Two Chicano Struggles in the Southwest* (Tucson: University of Arizona Press, 1994), 4, 24; Pulido, "Rethinking Environmental Racism: White Privilege and Urban Development in Southern California," *Annals of the Association of American Geographers* 90 (March 2000): 14–40. See also Paul Mohai, David Pellow, and J. Timmons Roberts, "Environmental Justice," *Annual Review of Environment and Resource* 34 (November 2009): 405–30.

6. Eileen McGurty, *Transforming Environmentalism: Warren County, PCBs, and the Origins of Environmental Justice* (New Brunswick, NJ: Rutgers University Press, 2007), 4; Robert D. Bullard, *Dumping in Dixie: Race, Class, and Environmental Quality* (Boulder, CO: Westview, 2000); Martin V. Melosi, "Equity, Eco-Racism, and Environmental History," in *Out of the Woods: Essays in Environmental History*, edited by Char Miller and Hal K. Rothman (Pittsburgh, PA: University of Pittsburgh Press, 1997), 194–211. Robert Gottlieb, *Forcing the Spring: The Transformation of the American Environmental Movement* (Washington, DC: Island Press, 2005), argues correctly that the antecedents of the environmental justice movement extend back to the Progressive Era, but much of the historical scholarship follows McGurty, continuing to focus on Warren County as the source of its modern impetus.

7. McGurty, *Transforming Environmentalism*, 7; Robert D. Bullard and Glenn S. Johnson, "Environmental Justice: Grassroots Activism and Its Impact on Public Policy Decision Making; Statistical Data Included," *Journal of Social Issues* 56 (Fall 2000).

8. Cortés to Burkett, May 27, 1994, 4–5; Cortés to Todd and Weisman, April 12, 2002.

9. On dating the origins of the local chapter of Cruz Azul, see *San Antonio Express*, September 17, 1921, 7; *La Prensa*, November 21, 1921, 5; Teresa Palomo Acosta, "Cruz Azul Mexicana," *Handbook of Texas Online*, accessed January 31, 2020, notes that the chapter's launch occurred in 1920; and Julie Leiniger Pycior,

Democratic Renewal and the Mutual Aid Legacy of U.S. Mexicans (College Station: Texas A&M University Press, 2014), 17–19, follows *La Prensa*'s lead. On the Mexican Consulate's $2,000 donation, see *San Antonio Express*, September 15, 1921, 11; *San Antonio Light*, September 17, 1921, 9. On the role of the Mexican consuls in local political affairs, see Gilbert G. González, *Mexican Consuls and Labor Organizing: Imperial Politics in the American Southwest* (Austin: University of Texas Press, 1999), 1–10, 27–36. Another framing of the work Cruz Azul dedicated itself to, according to *La Prensa*, was "Un sacrificio por el amor del género humano"; November 21, 1921, 5.

10. *La Prensa*, September 13, 1921, 1.

11. Ibid., 4.

12. Pycior, *Democratic Renewal*, 17–19; for a detailed exploration of nineteenth-century mutualists' extensive involvement with citywide celebrations, see Judith Berg Sobré, *San Antonio on Parade: Six Historic Festivals* (College Station: Texas A&M University Press, 2003),48–49, 73, 79–100.

13. Pycior, *Democratic Renewal*, 17–19.

14. *La Prensa*, November 21, 1921, 5.

15. Pycior, *Democratic Renewal*, 17–19. These discrepancies in economic opportunity held true for decades: Blackwelder, *Women of the Depression*, 75–108; Richard J. Harris, "Mexican American Occupational Attainments in San Antonio: Comparative Assessments," in Johnson et al., *Politics of San Antonio*, 52–71.

16. *La Prensa*, September 13, 1921, 1.

17. *La Prensa*, September 15, 1921, 8; *San Antonio Light*, September 16, 1921, 10.

18. "La tragedia de la inundación de San Antonio," 2nd ed., lists her name as Marta M. De Acosta; other sources, including the local English-language newspapers, give it as María De Costa; in 1922, local journalist María Luisa Garza would become the second president of Cruz Azul: Teresa Palomo Acosta, "Cruz Azul Mexicana," in *Handbook of Texas Online*, accessed July 26, 2020.

19. *San Antonio Light*, September 16, 1921, 10; *San Antonio Express*, September 17, 1921, 7; *La Prensa* offered daily updates on Cruz Azul's relief efforts as well as other mutualistas: see *La Prensa*, September 13, 1921, 1; September 14, 1921, 1–2, 8; September 15, 1921, 1, 8; September 16, 1921, 5; September 17, 1921, 6; September 18, 1921, 1, 12.

20. "El consulado recibio fondos," *La Prensa*, September 16, 1921, 5; "La recolección de fondos de la Sociedad Hidalgo," *La Prensa*, September 16, 1921, 1;

"Mexican Government Sends $2000 Fund for Natives' Relief," *San Antonio Light*, September 17, 1921, 9.

21. *San Antonio Light*, September 16, 1921, 10; "La tragedia," 50–52.

22. "La tragedia," 50; *San Antonio Light*, September 16, 1921, 10; *San Antonio Express*, September 17, 1921, 7. As these newspaper accounts also note, even the better-resourced Red Cross was having considerable difficulty distributing food and clothing, a result of what might be called disaster fatigue. The number of its volunteers began to decline, and fewer and fewer people with cars showed up to distribute supplies to the West Side relief shelters.

23. "La tragedia," 50.

24. Pycior, *Democratic Renewal*, 24–25; González, *Mexican Consuls*, 27–36.

25. Garza, *They Came to Toil*, 27–28; "An Act to Regulate Immigration," 47th Cong., Sess. 1, Chap. 376, 1882, loc.gov/law/help/statutes-at-large/47th-congress /session-1/c47s1ch376.pdf.

26. *La Epoca*, October 5, 1930, 3; October 12, 1930, 2: quoted in Julie Leininger Pycior, "La Raza Organizes: Mexican American Life in San Antonio, 1915–1930 as Reflected in Mutualista Activities" (PhD dissertation, University of Notre Dame, 1979), 225.

27. Quoted in Pycior, *Democratic Renewal*, 26–28.

28. Garza, *They Came to Toil*, 39; González, *Mexican Consuls*, 31–36.

29. Franklin D. Roosevelt, Second Inaugural Address, January 20, 1937, http:// historymatters.gmu.edu/d/5105/.

30. Robert B. Fairbanks, *The War on Slums in the Southwest: Public Housing and Slum Clearance in Texas, Arizona, and New Mexico, 1935–1965* (Philadelphia, PA: Temple University Press, 2014), 31–36; Audrey Granneberg, "Maury Maverick's San Antonio," *Survey Graphic* 1, no. 28 (July 1939): 423.

31. Pycior, *Democratic Renewal*, 27–28, 51, 80, 129.

32. Guadalupe San Miguel Jr., *"Let All of Them Take Heed": Mexican-Americans and the Campaign for Educational Equality in Texas, 1910–1981* (Austin: University of Texas Press, 1987), 69; Richard A. Garcia, *Rise of the Mexican American Middle Class: San Antonio, 1929–1941* (College Station: Texas A&M University Press, 1991), 253–300; Pycior, "La Raza Organizes," 226–32.

33. Pycior, *Democratic Renewal*, 124.

34. *Preliminary Report on a System of Major Streets*, 322–28.

35. *San Antonio Light*, May 6, 1935, 1; "Many Saved from Flood by Police, Firemen," *San Antonio Light*, May 10, 1935, 1, 14.

36. Breeding, "Flood of September 1946," 2–3; *San Antonio Express*, September 28,

1946.

37. "Rains Bring Heavy S.A. Damage," *San Antonio Light*, June 4, 1951, 1, 3.

38. "San Antonio Rain Routs 125 Families," *San Antonio Express*, June 4, 1951, 1, 3.

39. Walter H. Lilly and M. Winston Martin, *San Antonio Takes Stock and Looks Ahead: A Comprehensive Master Plan for San Antonio, Texas* (San Antonio: [n.p.], 1951), 192–207; Miller, *San Antonio: A Tricentennial History*, 126–29.

40. The Flood Control Act of 1954, Section 203, San Antonio Channel, reads: "The project for flood protection on the Guadalupe and San Antonio River, Texas, is hereby authorized substantially in accordance with the recommendations of the Chief of Engineers in House Document Numbered 344, Eighty-Third Congress at an estimated cost of $20,254,000." See Guadalupe and San Antonio Rivers, Texas, Chief of Engineers Report (February 1954), an Army Corps of Engineers report that served as the decision document for the authorized project, H. Exec. Doc. 344, 83rd Cong., 2nd Sess. This report concluded, in part, "that a serious flood problem exists within the city of San Antonio, an important military center and distribution point for a vast area in southwest Texas, and that a flood-protection project for this city to eliminate the flood menace is economically justified." Further, the report recommended "that a channel improvement project in San Antonio, Texas, be authorized at this time for construction by the Federal Government, substantially as outlined in this report, at an estimated first cost to the United States of $12,906,900."

41. "River Authority Plans Canal to Texas Coast," *San Antonio Express*, December 13, 1957, 8; "S.A. River Flood Project Underway," *San Antonio Light*, December 13, 1957, 28.

42. The Flood Control Act of 1954, Section 203: Guadalupe and San Antonio Rivers, Texas, Chief of Engineers Report (February 1954).

43. "S.A. River Flood Project Underway," 28.

44. "2 Reported Drowned," *San Antonio Light*, May 18, 1923, 5.

45. "Heavy Rains Return after S.A. Flooding," *San Antonio Express*, May 19, 1965, 1, 12A.

46. Ibid., 12A.

47. "Speed on Six Flood Projects Advocated," *San Antonio Express*, May 19, 1965, 10A.

48. Ibid.

49. Ronnie Dugger, "The Segregation Filibuster of 1957," in *Fifty Years of the*   ⊰ 229 ⊱
Texas Observer, edited by Char Miller (San Antonio, TX: Trinity University Press, 2004), 30.

50. "Henry B. González," History, Art and Archives, US House of Representatives, https://history.house.gov/People/Listing/G/GONZÁLEZ ,-Henry-B--(G000272)/ (accessed June 1, 2020); Miller, *Deep in the Heart*, 142–48.

51. "Gonzalez Due in S.A.," *San Antonio Light*, May 18, 1965, 1.

52. The San Antonio Texas Flood of May 1965: A Report, U.S. House Committee on Public Works, Miscellaneous Publications, 89th Congress (Washington, DC: GPO, 1965), 3–5.

53. Ibid., 4–6; Henry B. González, Testimony, "San Antonio Channel Project," Hearings, Reports and Prints of the House Committee on Appropriations, vol. 89, part 4, May 19, 1965, 1135–37.

54. His archival records reflect his watchdogging of SARA's projects: see Henry B. González Papers, 1946–2015, Briscoe Center for American History, the University of Texas at Austin, https://legacy.lib.utexas.edu/taro/utcah/00511 /cah-00511.html; see his correspondence to SARA in *A Guide to the San Antonio River Authority Records*, 1920–2016, University of Texas at San Antonio, https://legacy.lib .utexas.edu/taro/utsa/00272/utsa-00272.html.

55. Henry González, "From Participation to Equality," in *The Ethnic Moment: The Search for Equality in the American Experience*, edited by Philip L. Fetzer (Armonk, NY: M. E. Sharpe, 1996), 172.

56. *San Antonio River Authority, 1937–1987* (San Antonio River Authority, 1988), 11.

57. Rogers, *Cold Anger*, 111; Jan Jarboe, "Building a Movement: Mexican Americans Struggle for Municipal Services," *Civil Rights Digest*, Spring 1977, 39–46; Mark Warren, "A Theory of Organizing: From Alinsky to the Modern IAF," in *The Community Development Reader*, edited by James DeFillipis and Susan Saegert, 194–203 (New York: Routledge, 2008); Richard Buitron, *The Quest for Tejano Identity in San Antonio, Texas, 1913–2000* (New York: Routledge, 2004); Patrick J. Hayes, "COPS: Putting the Gospel into Action in San Antonio," in *Living the Catholic Social*

Tradition: Cases and Commentary, edited by Kathleen Maas Weigert and Alexia K. Kelly, 139–50 (Lanham, MD: Rowan and Littlefield, 2005).

58. Rogers, *Cold Anger*, 105–26.

59. Jarboe, "Building a Movement," 44.

60. Teresa Paloma Acosta and Ruthe Winegarten, *Las Tejanas: 300 Years of History* (Austin: University of Texas Press, 2004), 48.

61. Ibid.; Jan Jarboe Russell, "By Exposing Good Government League Becker Made History," *San Antonio Express-News*, June 26, 2006.

62. Nixon, *Slow Violence*, 3.

63. Jarboe, "By Exposing Good Government League"; Heywood Sanders, "Empty Taps, Missing Pipes: Water Policy and Politics," in Miller, *On the Border*, 141–68.

64. Cortés to Burkett, May 27, 1994, 3-4; Cortés to Todd and Weisman, April 12, 2002; Laura A. Wimberley, "Establishing 'Sole Source' Protection: The Edwards Aquifer and the Safe Drinking Water Act," in *On the Border*, 169–81; Robert Brischetto, Charles L. Cotrell, and R. Michael Stevens, "Conflict and Change in the Political Culture of San Antonio in the 1970s," in Johnson et al., *Politics of San Antonio*, 75–94; quote from Brinda Sarathy, "An Intersectional Reappraisal of the Environmental-Justice Movement," in Char Miller and Jeff Crane, eds., *The Nature of Hope: Grassroots Organizing, Environmental Justice, and Political Change* (Louisville: University Press of Colorado, 2018), 46.

65. Thomas A. Baylis, "Leadership Change in Contemporary San Antonio," in Johnson et al., *Politics of San Antonio*, 95–113; Tucker Gibson, "Mayoralty Politics in San Antonio, 1955–79," in Johnson et al., *Politics of San Antonio*, 114–29; Heywood Sanders, "Building a New Urban Infrastructure: The Creation of Postwar San Antonio," in *Urban Texas: Politics and Development*, 154–73; Rosales, *Illusion of Inclusion*, 83–158.

66. Garcia, *Rise of the Mexican American Middle Class*, 260.

67. Rogers, *Cold Anger*, 42–46; Mark A. Warren, "Building Democracy: Faith-Based Community Organizing Today," *Shelterforce Online*, January–February 2001; Char Miller, "Street Talk," in *A Tale of Two Cities: Atlanta and San Antonio. Proceedings of the 2005 Earl M. Lewis Symposium*, edited by Char Miller (San Antonio: Urban Studies Department, 2005), 4–13; Henry Cisneros, "People Make Change," in Miller, *Tale of Two Cities*, 127–43.

68. Rogers, *Cold Anger*, 125–26.

69. Ibid.; McGurty, *Transforming Environmentalism*, 9–11; Pulido, *Environmentalism and Economic Justice*, 13; Jarboe, "Building a Movement," 38–43; Moises Sandoval, "The Decolonization of a City," *Alicia Patterson Foundation Newsletter*, https:// web.archive.org/web/20061002100025/http://aliciapatterson.org/APF001977 /Sandoval/Sandoval04/Sandoval04.html.

CHAPTER 7: AFTERMATH

1. Vianna Davila, "Residents Cheer First Linear Park on West Side," *San Antonio Express-News*, November 20, 2010; some of the ideas in this chapter first appeared in different form in Char Miller, "Floods" and "Waterways," in *Three Hundred Years of San Antonio and Bexar County*, edited by Claudia R. Guerra (San Antonio, TX: Trinity University Press, 2018).

2. The Howard W. Peak Greenway Trails System, which, including the Museum and Mission Reaches of the San Antonio River, to date totals sixty-nine miles, sanantonio.gov/ParksAndRec/Parks-Facilities/Trails/Greenway-Trails.

3. Westside Creeks Restoration Project, westsidecreeks.com/about-the-creeks; Miller, *San Antonio: A Tricentennial History*.

4. SARA, Westside Creeks, Zarzamora Creek Trail, April 3, 2019, sara-tx .org/system/files/contracting_documents/2019-06/00141-8__Zarzamora _Landscape_plans_EX_8.pdf.

5. South Texas Floods, October 17–22, 1998, National Oceanic and Atmospheric Administration, US Department of Commerce, February 1999, 1–15; Floods in the Guadalupe and San Antonio River Texas, October 1998, USGS, https://pubs.usgs.gov/fs/FS-147-99/; Sig Christenson and Adolfo Pesquera, "3 Killed as Floods Hit S.A. Area," *San Antonio Express-News*, October 18, 1998, 1A; Adolfo Pesquera, "Bad Weather Forces Residents, Rescuers to Evacuate," *San Antonio Express-News*, October 18, 1998, 16A; photos: "Record Storm," *San Antonio Express-News*, October 18, 1998, 17A.

6. Ralph Winingham, "Tunnel a Success in Debut," *San Antonio Express-News*, October 19, 1998, 1A.

7. Rick Casey, "Tale of a Basement in King William," *San Antonio Express-News*, October 19, 1998, 3A.

8. City of San Antonio Flood Mitigation Plan, December 2000, appendix K and W, twdb.texas.gov/publications/reports/contracted_reports/doc/2000001011.pdf.

9. Gumprecht, *Los Angeles River*; Andrew W. Honker, "'A Terrible Calamity Has Fallen upon Phoenix': The 1891 Flood and Salt River Valley Reclamation," *Journal of Arizona History* 43, no. 2 (2002): 109–32; Andrew C. Revkin, "Peeling Back Pavement to Expose Watery Havens," *New York Times*, July 16, 2009.

10. Elaine Ayala, "San Juan Acequia Flows Again," *San Antonio Express-News*, September 28, 2011.

11. San Antonio River Improvements Project, Mission Reach, sanantonioriver.org/mission_reach/mission_reach.php.

12. San Antonio River Improvements Project, Museum Reach, sanantonioriver.org/museum_reach/museum_reach.php.

13. "Confluence Park: A Confluence of Watershed and Conservation," *Modern in SA*, March 11, 2019; Tenna Florian to Char Miller, email communication, March 16, 2020.

14. San Pedro Creek Culture Park, https://spcculturepark.com/about/; for background on the San Pedro Creek Tunnel, see sariverauthority.org/services/flood-management/engineering-projects/san-pedro-creek-tunnel.

-{ ACKNOWLEDGMENTS }-

Sometime in the late 1990s, while I was grading student papers in my light-flooded office at Trinity University in San Antonio, a man walked up to my open door, knocked on the metallic frame, and asked if we could talk. Even as I was saying yes, I spotted the grocery bag he was carrying. Emblazoned with the three-letter logo of the city's major supermarket chain, the brown paper sack looked heavy. A better historian would have remembered his name, but I confess my eyes kept flicking to what he was hefting. He introduced himself as a friend of my colleague Aaron Konstam, who had told him that I was working on a history of the 1921 flood and its impact on San Antonio. "If you are," he said, "you are going to want to see this document."

It was more than a document. It was a bulky compendium. He pulled it gently out of the bag and began to turn the stiff pages of the US Army report that contained a full description of the catastrophe and details pertaining to its search-and-rescue operations; vivid photographs that the Signal Corps took at first light on September 10, 1921, as floodwaters raced down city streets; and additional images of damaged bridges, neighborhoods, and streetscapes. Best of all were the aerial mosaics that pilots flying out of Kelly Airfield had snapped of the devastation along three of the city's main watersheds. I was speechless. And grateful: my visitor indicated that he worked as a civilian on Fort Sam Houston, saw that the report was going to be tossed out, salvaged it, and in time learned of my research. *West Side Rising* owes an enormous debt to his preservationist impulse.

I am every bit as indebted to those who attended my local talks about the flood and its aftermath and who shared their families' stories afterward or later wrote me about their grandparents' or parents' harrowing experiences. I hope I have done justice to these shared remembrances.

As for the many friends and colleagues who have listened to my flood-related stories over the past twenty-five years or so, thanks for your patience! At Trinity University, where I taught for many years, the list is long. Conversations with historians Gary Kates, John Martin, Alida Metcalf, and Linda Salvucci proved heuristic. So were those with my colleagues in Urban Studies, especially Earl Lewis, Cathy Powell, Heywood Sanders, and Christine Drennon; and John Donahue and Richard Reed. John Booth was another great sounding board. Library directors at Trinity, from Richard Werking and Diane Graves to Chris Nolan, have been amazing resources, as were (and are) their colleagues. Early collaborations with David Johnson and Sanders, and research help from Kay Reamey and Stephanie Hetos Coxe, gave the project its initial boost. In the late 1980s Donald E. Everett handed me a trove of material on Olmos Park and the Olmos Dam, documents that helped me see the city with fresh eyes. My perspectives have continued to evolve because of the vision of architects David Lake and Ted Flato; Tenna Florian and Denise De Leon of Lake/Flato provided a deeper understanding of Confluence Park and offered some great images. Friends and neighbors David Ladensohn and Buddy Gardner have kept me abreast of ongoing flooding issues in the city. Lewis Fisher and Maria Watson Pfeiffer's unsurpassed knowledge of San Antonio has been indispensable to my work and to that of so many others. I owe an overdue shout-out to the members of the Open Space Advisory Board and the Tree Preservation Ordinance Committee who taught me about local watersheds, linear parks, and urban tree canopies. Long morning walks through Olmos Park with Nutmeg were enlightening about how that community was put together.

Archivists at the National Archives, University of California–Berkeley, and the University of California–Riverside, like those at the San

Antonio River Authority, Trinity University, the University of Texas at San Antonio, and the Claremont Colleges, have shared relevant primary sources—documents, images, and ephemera—that brought the research to life. Special thanks to Casey Dunn for permission to use photographs of Confluence Park. I have published some of my ideas about the 1921 flood's significance in the *Texas Observer* and the *Southwestern Historical Quarterly*, and I am grateful to Lou Dubose and Ryan Schumacher for their editorial encouragement and red pens. As engaging has been the chance not once but twice to speak about the city's complex flood control politics at the University of Colorado Boulder; I am grateful to Patty Limerick, Kurt Gutjahr, and Phoebe Young for the invitations to visit their beautiful campus. Julie Leininger Pycior of Manhattan College generously shared her archive of material on Cruz Azul. My understanding of the nuances of environmental justice has been enriched by conversations with my many colleagues in Claremont who teach and write about this field: Aimee Bahng, Guillermo Douglass-Jaimes, Zayn Kassam, Joanne Nucho, Brinda Sarathy, Heather Williams, and Rick Worthington (not to say, our many shared students).

The initial stages of research for this project were underwritten by several Faculty Development Grants from Trinity University; timely leaves from there and from Pomona College facilitated different parts of the research and writing process. Leigh Lieberman and Aaron Hodges in the Digital Humanities at the Claremont Colleges provided generous funding for underwriting the gifted translation by José Emilio Bencosme Zayas, founder of Adben S.R.L., of "La tragedia de la inundación de San Antonio," which made an essential contribution to *West Side Rising*. Digital Humanities also supported the stellar work of student researchers that have transformed *West Side Rising*. Led by Sarah Osailan of Claremont Graduate University, Anam Mehta, a member of the Pomona College class of 2021, along with Scripps College graduates Katie Graham and Natalie Quek, spent hours bringing the US Army report's photographic evidence into the digital age and have developed accompanying story maps. Nicole Arce, who graduated from Pomona in 2021, provided critical translations of key articles in *La Prensa*. Like so many other students

{ 235 }

over the years, this quintet has pushed me to think beyond what I know to imagine innovative ways of telling oft-told stories.

That is why, in turn, I have been fortunate to collaborate so frequently with Trinity University Press. Poet Barbara Ras, founding director of the reinvigorated press, encouraged me to experiment with literary form and narrative voice, gave me opportunities I could not have envisioned on my own, and then deftly edited my tangled prose. Tom Payton, her successor, has been beyond supportive, ever alert to new initiatives. Margy Avery, as point person on *West Side Rising*, has nudged the book into shape, thought creatively about its physical form and design, and secured two anonymous readers for the manuscript who offered an array of constructive suggestions. Over the years she has just as adeptly taken me to some cool new restaurants in San Antonio, once a foodie desert of sorts. It makes my day to get an email from Trinity's Sarah Nawrocki, Burgin Streetman, LeeAnn Sparks, or Stephanie Mortis Stevens, because that's when good things happen.

Even better things occur when our family gets together. Judi isn't just the best editor ever, she's our family's centripetal force—with Sam and Campbell, Ben and Caitlin, and Rebecca and Sean bringing their own magnetic energy. I cannot imagine my life without them and happily dedicate *West Side Rising* to them.

CREDITS

Division, Farm Security Administration / Office of War Information Collection.

Page 101   Letters and Photographs Pertaining to Construction of Olmos Dam, Texas. Reproduced by permission from the University of California–Riverside Special Collections and University Archives.

Page 108   General Photograph Collection, MS 362. Reproduced by permission from the University of Texas at San Antonio Libraries Special Collections.

Page 124   Reproduced by permission from Edwards Aquifer online collection.

Page 143   Reproduced courtesy of Keystone View Company.

Page 144   General Photograph Collection, 101-362. Reproduced by permisssion from the University of Texas at San Antonio Special Collections.

Page 146   Map drawn by Anam Mehta. Collection of the author.

Page 156   Reproduced by permission from the University of Texas at San Antonio Special Collections.

Page 166   *San Antonio Light*, San Antonio River Authority Records: The 1935 Flood, Box 48. Reproduced by permission from the University of Texas at San Antonio Libraries Special Collections.

Page 167   *San Antonio Light* Photograph Collection, MS 359. Reproduced by permission from the University of Texas at San Antonio Libraries Special Collections.

Page 175   San Antonio River Authority Records, 1920–2018. University of Texas at San Antonio Libraries Special Collections.

Pages 191, 192   Photographs by Casey Dunn. Courtesy of the San Antonio River Foundation.

{ INDEX }

Note: page numbers in italics refer to figures; those followed by *m* refer to maps; those followed by "n" refer to notes, with note number.

CHAR MILLER is the W. M. Keck Professor of Environmental Analysis and History at Pomona College. He is the editor of *On the Border: An Environmental History of San Antonio* and *Fifty Years of the Texas Observer* and the author of *Gifford Pinchot and the Making of Modern Environmentalism, Not So Golden State: Sustainability vs. the California Dream, On the Edge: Water, Immigration, and Politics in the Southwest, Deep in the Heart of San Antonio: Land and Life in South Texas, San Antonio: A Tricentennial History,* and *Public Lands, Public Debates: A Century of Controversy.* Miller is a frequent contributor to print, electronic, and social media.

JULIÁN CASTRO, a lawyer and politician from San Antonio, Texas, launched his public service career in 2001, becoming the youngest councilman in the city's history. He served as mayor from 2009 to 2014, and as US secretary of housing and urban development from 2014 to 2017. He is the author of *An Unlikely Journey: Waking Up from My American Dream.*